Science, Technology, and Society

The Impact of Science
in the
19th Century

Science, Technology, and Society

The Impact of Science in the 19th Century

David E. Newton, Neil Schlager, Kelle Sisung, Editors

VOLUME 2

Mathematics

Physical Science

Technology and Invention

AN IMPRINT OF THE GALE GROUP

DETROIT · NEW YORK · SAN FRANCISCO
LONDON · BOSTON · WOODBRIDGE, CT

Science, Technology, and Society

The Impact of Science in the 19th Century

David E. Newton, Neil Schlager, Kelle Sisung, *Editors*
Stephanie Islane Dionne, *Contributing Editor*

STAFF

Christine Slovey, *U•X•L Senior Editor*
Elizabeth Grunow, *U•X•L Editor*
Gerda-Ann Raffaelle, *U•X•L Editor*
Carol DeKane Nagel, *U•X•L Managing Editor*
Tom Romig, *U•X•L Publisher*

Rita Wimberley, *Senior Buyer*
Evi Seoud, *Assistant Production Manager*

Debra J. Freitas, *Permissions Associate*
Tracey Rowens, *Senior Art Director*
Robyn V. Young, *Image Editor*
Robert Duncan, *Imaging Specialist*

Library of Congress Card Number: 2001089546

This publication is a creative work copyrighted by U•X•L and fully protected by all applicable copyright laws, as well as by misappropriation, trade secret, unfair competition, and other applicable laws. The author and editors of this work have added value to the underlying factual material herein through one or more of the following: unique and original selection, coordination, expression, arrangement, and classification of the information. All rights to this publication will be vigorously defended.

Front cover photographs reproduced by permission of The Bettmann Archives and The Granger Collection.

Copyright © 2001
U•X•L, an imprint of the Gale Group

All rights reserved, including the right of reproduction in whole or in part in any form.

ISBN 0-7876-4874-4 (set)
ISBN 0-7876-4875-2 (vol. 1)
ISBN 0-7876-4876-0 (vol. 2)

Printed in the United States of America
10 9 8 7 6 5 4 3 2 1

Contents

Reader's Guide . . . . *xi*
Advisory Board . . . . *xiii*
Chronology . . . . *xv*
Words to Know . . . . *xxiii*

VOLUME 1

chapter one Life Science

CHAPTER CONTENTS 1

CHRONOLOGY 2

OVERVIEW 3

ESSAYS
Johann Blumenbach and the Classification
of Human Races 7
Advances in Cell Theory 10
Birth of the Agricultural Sciences 16
Gregor Mendel Discovers the Basic Laws
of Heredity 21
Thomas Malthus and the Study
of Population 25
Charles Darwin's Theory of Evolution 29
Social Darwinism 36
Louis Pasteur and the Germ Theory 39
The Popularization of Science 45

Contents

Neanderthals and the Search for
Human Ancestors 48

BIOGRAPHIES
John James Laforest Audubon 54
Spencer Fullerton Baird 56
Ernst Haeckel 58
Thomas Henry Huxley 59
Ida Henrietta Hyde 61
Matthew Fontaine Maury 62
Lorenz Oken 64
Alfred Russel Wallace 65

BRIEF BIOGRAPHIES 67

RESEARCH AND ACTIVITY IDEAS 74

FOR MORE INFORMATION 75

chapter two Medicine

CHAPTER CONTENTS 79

CHRONOLOGY 80

OVERVIEW 81

ESSAYS
The Development of New Systems of
Alternative Medicine 84
Advances in Surgical Techniques 90
The Battle against Tuberculosis 94
Nineteenth Century Biological Theories
about Race 98
Medical Beliefs about Women 100
Establishment of Schools for Individuals
with Special Needs 103
Nineteenth-Century Advances
in Dentistry 106
Advances in the Treatment of
Mental Illness 109
Developments in Military Medicine 113
The Rise of Tropical Medicine 116
Creation of the Public Health System 120
Koch's Postulates Are Used to
Identify Diseases 124

Contents

BIOGRAPHIES
Clara Barton 128
William Beaumont 130
Elizabeth Blackwell 132
Paul Ehrlich 135
Sigmund Freud 137
Shibasaburo Kitasato 139
Charles Louis Alphonse Laveran 140
Florence Nightingale 141
William Osler 143
Philippe Pinel 144
Wilhelm Konrad Roentgen 145
Ignaz Phillip Semmelweis 147

BRIEF BIOGRAPHIES 149

RESEARCH AND ACTIVITY IDEAS 157

FOR MORE INFORMATION 158

INDEX xxxix

VOLUME 2

chapter three — Mathematics

CHAPTER CONTENTS 163
CHRONOLOGY 164
OVERVIEW 165

ESSAYS
Early Attempts to Formulate a Mathematical
Theory of Electricity 168
Developments in Mathematics Education 172
Advances in the Mathematics of Logic 175
The Rise of Probability Theory 179

BIOGRAPHIES
Charles Babbage 182
George Boole 185
Augustin-Louis Cauchy 186
Jean-Baptiste Joseph Fourier 187
Evariste Galois 189
Carl Friedrich Gauss 190

Contents

Sophie Germain **193**
Felix Klein **194**
Pierre-Simon Laplace **195**
Nikolai Ivanovich Lobachevsky **197**
Jules-Henri Poincaré **199**

BRIEF BIOGRAPHIES **200**

RESEARCH AND ACTIVITY IDEAS **207**

FOR MORE INFORMATION **208**

chapter four Physical Science

CHAPTER CONTENTS **211**

CHRONOLOGY **212**

OVERVIEW **213**

ESSAYS
Development of the Concept of Energy **216**
Nineteenth-Century Advances in
Electromagnetic Theory **221**
Finding Order among the Elements: The
Periodic Table **226**
Lord Rayleigh's Theory of Sound **232**
Discovery of Radioactivity **236**
Discovery of the Inert Gases **243**
Heinrich Hertz Discovers Radio Waves **247**
Charles Lyell Popularizes the Principle
of Uniformitarianism **251**
Developments in Electrochemistry **255**
Discovery of Ice Ages **261**

BIOGRAPHIES
Jean Louis Rodolphe Agassiz **265**
Friedrich Wilhelm Bessel **267**
Marie Curie **269**
John Dalton **271**
Michael Faraday **273**
Joseph Louis Gay-Lussac **275**
Josiah Willard Gibbs **277**
Joseph Henry **279**
Caroline Lucretia Herschel **280**
Charles Lyell **282**
James Clerk Maxwell **283**

BRIEF BIOGRAPHIES 285	Contents
RESEARCH AND ACTIVITY IDEAS 293	
FOR MORE INFORMATION 294	

chapter five

Technology and Invention

CHAPTER CONTENTS 297
CHRONOLOGY 298
OVERVIEW 299

ESSAYS

The Horseless Carriage: Invention of the First
Automobiles 303
Revolutions in the Publishing Industry 308
Instant Messaging: The Invention of
the Telegraph 313
Alexander Graham Bell Invents
the Telephone 316
Capturing Life on Screen: The Invention of
Motion Pictures 320
Electricity Powers the
Nineteenth Century 324
The Steam-Powered Locomotive
Transforms Transportation 331
The Mechanization of Textile Weaving 336
Progress in Food Preservation 340
Advances in the Chemistry of Explosives 344
Charles Goodyear Invents the Process
of Vulcanization 348

BIOGRAPHIES

Nicolas Appert 352
Alexander Graham Bell 353
Karl Friedrich Benz 355
Henry Bessemer 356
Augusta Ada Byron, Countess
of Lovelace 358
Thomas Alva Edison 359
Margaret E. Knight 361
Cyrus Hall McCormick 362
Norbert Rillieux 363
Isaac Merrit Singer 365

Contents

BRIEF BIOGRAPHIES 366
RESEARCH AND ACTIVITY IDEAS 372
FOR MORE INFORMATION 373
INDEX xxxix

Reader's Guide

Science and society experienced tremendous changes in the nineteenth century. At the beginning of the century people traveled by horse, or horse-drawn carriage, and communication outside of one's own town or city took days, weeks, or months. Scientists had very little understanding of physiology and anatomy; blood-sucking leaches were a popular medical treatment. By the end of the century inventions such as the telephone, radio, electric light, and the steam engine changed the way people lived, worked, and played. Discoveries in the fields of physics, chemistry, and biology changed the way people viewed the natural world and humankind's place in it. Developments in the field of medicine made treatments safer and more effective. *Science, Technology, and Society: The Impact of Science in the 19th Century* is designed to help students understand the impact that nineteenth-century science had on the course of human history. Scientific discoveries and developments are examined within their historical context, showing how social trends and events influenced science and how scientific developments changed people's lives.

Format

Science, Technology, and Society: The Impact of Science in the 19th Century is divided into five chapters across two volumes. The Life Science and Medicine chapters appear in Volume One. The Mathematics, Physical Science, and Technology and Invention chapters appear in Volume Two. The following sections appear in each chapter:

Chronology: A timeline of key events within the chapter's discipline.

Overview: A summary of the scientific discoveries and developments, trends, and issues within the discipline.

Essays: Topic essays describing major discoveries and developments within the discipline and relating them to social history. Information in

Reader's Guide

the topic essays is divided under the standard rubrics Overview, Background, and Impact.

Biographies: Biographical profiles providing personal background on important individuals within the discipline, and often introducing students to additional important issues in science and society in the nineteenth century.

Brief Biographies: Brief biographical mentions introducing students to the major accomplishments of other notable scientists, researchers, teachers, and inventors important within the discipline.

Research and Activity Ideas: Offering students ideas for reports, presentations, or classroom activities related to the topics discussed in the chapter.

For More Information: Providing sources for further research on the topics and individuals discussed in the chapter.

Other features

Sidebars in every chapter highlight interesting events, issues, or individuals related to the subject. More than 130 black-and-white photographs help illustrate the discoveries and the individuals who made them. In addition, cross-references to subjects discussed in other topic essays are indicated with "see references" in parenthesis while cross-references to individuals discussed elsewhere in the title are indicated by boldface type and "see references" in parenthesis. Each volume concludes with a cumulative subject index so that students can easily locate the people, places, and events discussed throughout *Science, Technology, and Society: The Impact of Science in the 19th Century*.

Comments and Suggestions

We welcome your comments on *Science, Technology, and Society: The Impact of Science in the 19th Century* and suggestions for other science topics to consider. Please write: Editors, *Science, Technology, and Society: The Impact of Science in the 19th Century*, U•X•L, 27500 Drake Rd., Farmington Hills, Michigan 48331-3535; call toll-free: 1-800-877-4253; fax to (248) 414-5043; or send e-mail via http://www.galegroup.com.

Advisory Board

Special thanks are due to U•X•L's *Science, Technology, and Society* advisors. The following teachers, librarians, and media specialists offered invaluable comments and suggestions when this work was in its formative stages:

Dr. Josesph L. Hoffman
Director of Technology
West Bloomfield School District
West Bloomfield, Michigan

Dean Sousanis
Science Chairman
Almont High School
Almont, Michigan

Leila J. Sprince
Head of Youth Services
Broward County Library
Ft. Lauderdale, Florida

Chronology

1800 Italian physicist Alessandro Giuseppe Volta invents the first battery.

1800 English chemist Humphry Davy introduces the first chemical anesthetic, nitrous oxide or laughing gas.

1801 French inventor Joseph-Marie Jacquard invents a mechanical loom that operates according to instructions stored on punch cards.

1801 German mathematician Johann Karl Friedrich Gauss publishes *Disqusitiones Arithmeticae* (Arithmetical Researches), which greatly expands number theory.

1802 French biologist Jean-Baptiste de Lamarck and German naturalist Gottfried Treviranus simultaneously coin the term *biology*, to refer to the study of all living bodies.

1803 English chemist John Dalton proposes the first modern atomic theory.

1803 The first steam engine is invented by English engineer Richard Trevithick.

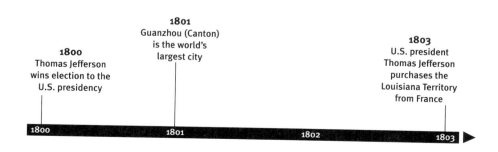

Chronology

1804 French inventor Nicolas Appert develops the process of canning to preserve food, and opens the first canning factory.

1805 German chemist Friedrich Wilhelm Adam Serturner isolates morphine, which is to become one of the most popular pain relievers in history.

1807 American inventor Robert Fulton builds the *Clermont,* the first steamboat to operate successfully on a commercial basis.

1809 Lamarck publishes *Zoological Philosophy.* The book offers a theory of evolution saying that changes in body structure are passed down from one generation to the next.

1810 The first privately established mathematics journal is founded at the École Polytechnique in France by Joseph Diaz Gergonne. It is called the *Annals of Pure and Applied Mathematics.*

1812 The first automated printing press is designed by German inventor Friedrich König.

1812 French naturalist Georges Cuvier proposes the theory of catastrophism, which says that the world has been swept by a number of major catastrophes in its history. Each catastrophe, he says, wipes out all living organisms, which are then replaced by plants and animals of entirely new kinds.

1812 French astronomer and mathematician Pierre-Simon Laplace publishes *Théorie Analytique des Probabilités,* and establishes the basis for the modern science of probability.

1816 French physician René-Théophile-Hyacinthe Laënnec invents the stethoscope.

1820 Danish physicist Hans Christian Oersted finds that an electric current can produce a magnetic field. The discovery forms the basis of many modern electromagnetic devices.

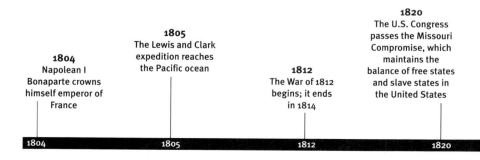

1804 Napolean I Bonaparte crowns himself emperor of France

1805 The Lewis and Clark expedition reaches the Pacific ocean

1812 The War of 1812 begins; it ends in 1814

1820 The U.S. Congress passes the Missouri Compromise, which maintains the balance of free states and slave states in the United States

Chronology

1822 French mathematician Jean-Baptiste Joseph Fourier discovers a method for analyzing periodic motion. Eventually known as Fourier series, it proves invaluable to the study of all types of waves.

1823 English chemist William Prout shows that hydrochloric acid is a component of gastric juices.

1825 American army surgeon William Beaumont begins a long series of investigations on the process of digestion by studying an open wound in the stomach of a young French Canadian man, Alexis St. Martin.

1826 French inventor Joseph Niepce produces the first permanent photograph.

1826 German mathematician August Leopold Crelle launches the *Journal for Pure and Applied Mathematics.* Known as *Crelle's Journal,* it is the first periodical to focus exclusively on research in pure mathematics.

1827–1838 American naturalist and artist John James Audubon publishes *Birds of America.*

1828 English mathematician George Green publishes his *Essay on the Application of Mathematical Analysis to the Theories of Electricity and Magnetism.* It eventually provides the basis for understanding electric and magnetic fields.

1829 French mathematician Evariste Galois presents a paper introducing group theory to the French Academy of Sciences. He presents a second version in 1830. The papers are mysteriously lost by members of the academy, but eventually Galois's work resurfaces and his theories are published in 1846.

1831 English chemist Michael Faraday and American physicist Joseph Henry invent the first electric generators and electric motors.

1831 Scottish botanist Robert Brown discovers the nucleus of a cell.

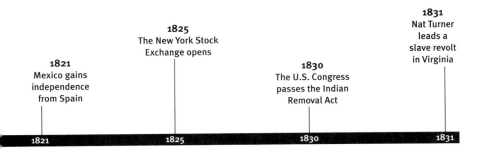

The Impact of Science in the 19th Century ■ Volume Two

Chronology

1831 American inventor Cyrus Hall McCormick invents his mechanical reaping machine.

1836 German physiologist Theodor Schwann discovers the role of pepsin in digestion.

1837 Swiss American geologist Jean Agassiz suggests that Earth has been covered a number of times in the past by massive ice sheets, a theory that would not be accepted for some twenty-five years.

1838 German botanist Matthias Jakob Schleiden announces that plant tissue is composed of cells.

1838 British astronomer Thomas Henderson makes the first determination of the distance to a star, Alpha Centauri. At about the same time, German astronomer Friedrich Wilhelm Bessel measures the distance to another star, 61 Cygni.

1839 American inventor Charles Goodyear accidentally spills a mixture of rubber and sulfur on his wife's stove, discovering the process known as vulcanization.

1839 German physiologist Theodor Schwann suggests that all animal tissue is made up of cells. Schwann's ideas are combined with those of Matthias Jakob Schleiden to form cell theory.

1844 Having patented the telegraph in 1837, American inventor Samuel F. B. Morse successfully transmits the first Morse code message over a telegraph circuit between Baltimore and Washington: "What hath God wrought?"

1846 French astronomer Urbain-Jean-Joseph Leverrier and English astronomer John Couch Adams discover the planet Neptune at almost the same time.

1847 English mathematician George Boole writes the first paper to suggest that logical ideas can be expressed using mathematical symbols; seven years later he fully develops his theories, which are now known as Boolean algebra.

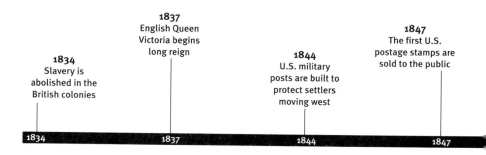

1834 Slavery is abolished in the British colonies

1837 English Queen Victoria begins long reign

1844 U.S. military posts are built to protect settlers moving west

1847 The first U.S. postage stamps are sold to the public

Chronology

1847 Hungarian physician Ignaz Phillip Semmelweis enforces a rule that doctors at the hospital in his charge wash their hands before and after touching a patient.

1854 While caring for British patients in the Crimean War, Florence Nightingale establishes new standards for nursing and patient care.

1854 English physician John Snow discovers that more than five hundred cases of cholera in London occur within a few blocks of a public water pump that draws water from the area around a sewer pipe. He has the pump handle removed, and the epidemic disappears. Snow's work marks the beginning of public health practices in disease prevention.

1854 German mathematician Georg Friedrich Bernhard Riemann proves that various types of non-Euclidian geometry are possible.

1856 English engineer and inventor Henry Bessemer announces his invention of an efficient and inexpensive new method for producing steel.

1856 The skeletal remains of a primitive human are found in the Neander River Valley in Germany. The specimen is later given the name Neanderthal Man.

1859 Charles Darwin's *On the Origin of Species* outlines the modern theory of evolution by natural selection.

1862 French chemist Louis Pasteur announces the germ theory of disease, one of the most important milestones in the creation of modern medicine.

1865 Laying the groundwork for antiseptic surgery, English surgeon Joseph Lister uses phenol (also called carbolic acid) to prevent infection during an operation on a compound fracture.

1865 The London Mathematical Society is formed.

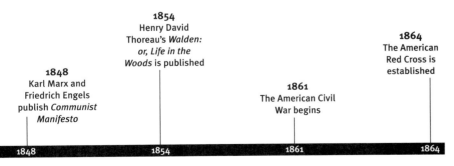

Chronology

1865 German mathematician August Ferdinand Möbius invents the Möbius strip, a figure that has one side and one edge. The discovery contributes to the establishment of a new branch of mathematics known as topology (the mathematical study of surfaces).

1865 Austrian monk Gregor Johann Mendel publishes his theories of heredity.

1865 English physicist James Clerk Maxwell develops a set of mathematical equations that shows the relationship between electrical and magnetic fields.

1869 Russian chemist Dmitri Ivanovich Mendeleyev develops the first modern version of the periodic table.

1871 Darwin publishes *The Descent of Man*, which discusses how humans fit into the evolutionary process.

1872 German botanist Ferdinand Julius Cohn publishes a three-volume work that marks the beginning of bacteriology as a distinct field of science.

1876 German inventor Karl Paul Gottfried von Linde builds the first practical refrigerator.

1876 American inventor Alexander Graham Bell invents the telephone.

1877 American inventor Thomas Alva Edison builds the first phonograph.

1879 English inventor Joseph Wilson Swan and Edison simultaneously produce the first practical incandescent light bulb.

1881 French chemist Louis Pasteur develops the first artificially created vaccine and successfully vaccinates sheep against anthrax, a deadly disease that affects animals and humans.

1882 The discovery of chromosomes and cell mitosis (cell division) is reported in *Cell Substance, Nucleus, and Cell Division*, written by German anatomist Walther Fleming.

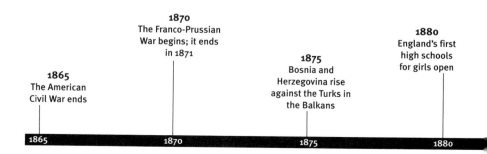

Chronology

1885 German engineer Karl Friedrich Benz builds what may be considered the first true automobile, a vehicle which is powered by a gasoline-burning internal combustion engine.

1885 Pasteur successfully treats a boy infected with rabies by injecting him with a weakened sample of rabies virus.

1887 American physicist Albert Michelson and American chemist Edward Morley carry out one of the most famous experiments in scientific history. The experiment fails to detect the presence of an ether, long thought by scientists to permeate the universe.

1888 The first professional mathematics society in the United States is created. Originally called the New York Mathematical Society, it becomes the American Mathematical Society in 1894.

1888 German physicist Rudolph Hertz first produces and detects radio waves.

1890 German bacteriologist Emil Adolf von Behring develops an antitoxin vaccine that protects against diphtheria, an accomplishment for which he was awarded the first Nobel Prize for physiology or medicine in 1901.

1890 American inventor Herman Hollerith builds the first mechanical calculator using principles on which modern computers are based.

1894 British scientists John William Strutt (Lord Rayleigh) and William Ramsay isolate argon, the first in a series of inert gases.

1895 German physicist Wilhelm Conrad Roentgen discovers X rays.

1895 Auguste and Louis Lumière invent the cinematograph, a portable hand-cranked camera that can shoot, print, and project motion pictures. Historians date this year as the birth of the motion picture.

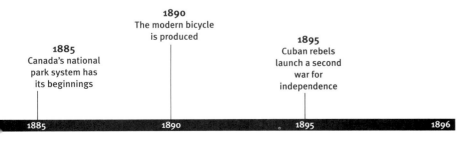

Chronology

1896 French physicist Antoine-Henri Becquerel finds that certain materials give off X ray-like radiation, a process later named radioactivity.

1896 Dutch physician Christiaan Eijkman shows that the disorder known as beriberi is caused by a dietary deficiency, the first such condition to be identified.

1898 Dutch botanist Martinus Willem Beijerinck searches for the cause of tobacco mosaic disease. He discovers an agent even smaller than bacteria and calls it a virus.

1899 The German pharmaceutical firm of Farbenfabriken Bayer introduces aspirin.

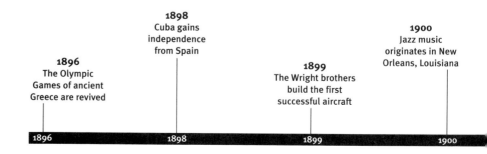

1896 The Olympic Games of ancient Greece are revived

1898 Cuba gains independence from Spain

1899 The Wright brothers build the first successful aircraft

1900 Jazz music originates in New Orleans, Louisiana

Words to Know

A

acoustics The science dealing with sound.

acupuncture A very old method of treatment that originated in China in which disorders of the body are treated by inserting long needles into certain points in the body.

adaptation The process by which organisms or groups of organisms become adjusted to their environment.

agricultural chemistry The field of science in which chemicals are used to increase the quality or amount of farm crops produced. Synthetic fertilizers, pesticides, and herbicides (weed-killers) are examples of such chemicals.

air A mixture of nitrogen, oxygen, argon, carbon dioxide, and other gases that make up Earth's atmosphere; in ancient times, the name used for a gas.

alternating current (AC) An electrical current that changes the direction in which it flows many times per second.

amalgam An alloy (mixture) of mercury and at least one other metal, used to fill cavities in teeth.

analgesia A chemical that offers pain relief.

anesthetic Any chemical that will cause the loss of consciousness and immunity to pain.

anthrax An infectious disease that occurs most commonly among cattle and sheep, but that can also be transmitted to humans.

anthropology The study of the origin of man.

antiseptic A chemical that kills disease-causing microorganisms.

Words to Know

arithmetic The branch of mathematics that involves calculations, such as addition, subtraction, multiplication, and division, with real numbers.

art therapy A form of treatment in which patients are encouraged to express their thoughts and feelings through painting, sculpture, or some other form of art.

assembly line A series of devices, materials, and workers that allow one object after another to be assembled step-by-step.

atomic weight The weight of an atom expressed in atomic mass units (amu). An atomic mass unit is defined as $1/12$ the mass of a carbon-12 atom.

B

bacillus anthracis The microbe that causes anthrax.

bacteria A group of one-celled organisms.

battery A device that uses chemical changes to generate an electric current.

binary system A method of counting that uses only two digits, 1 and 0.

blood-letting A method doctors once used for treating medical disorders in which blood is drawn from a patient's body; it was believed this would correct imbalances in the blood that caused disease.

Braille system A system that allows visually impaired and sightless people to read by using raised dots on a page to represent letters, numbers, and symbols.

bridge Another name for a denture; usually refers to a partial denture that is anchored on each end to real teeth.

C

calculus A field of mathematics that deals with the rate at which things change.

canning A method for preserving food by cooking it and sealing it in air-tight containers.

carbolic acid Another name for phenol, a compound that can be used as a disinfectant.

catastrophism A geological theory that says Earth as we know it today was produced by a series of sudden and spectacular changes by processes that can no longer be observed today.

cell The building block of all living things.

celluloid A hard, transparent filmlike material from which motion picture films were once made.

Words to Know

cephalic index A method for describing the shape of a human skull. It is the ratio of skull length to skull width.

chassis The supporting frame, or "body," of a vehicle.

chemical element A substance that can not be reduced to any simpler form of matter.

chiropractic A field of alternative medicine based on the philosophy that disease can be prevented by making sure that the body's main systems are kept in proper working order, primarily by making necessary spinal adjustments.

cholera An infectious disease that causes nausea, vomiting, and dehydration; symptoms develop quickly and are severe.

chromosome A coiled structure in the nucleus of a cell that carries the cell's deoxyribonucleic acid (DNA).

classification A method of organizing plants and animals into categories based on their appearance and the natural relationships between them.

climate The trend in average weather conditions over a relatively long period of time, usually at least thirty years.

coal gas A type of fuel produced by heating coal in the absence of air.

cochlea A bony mass in the inner ear that vibrates when struck by sound waves.

consumption A name that was used for tuberculosis in the nineteenth century.

cordite A type of explosive made by combining nitroglycerine, guncotton, and vaseline.

cosmology The science dealing with the birth and evolution of the universe.

cottage industry A business operated, usually in a private home, by an individual, a single family, or group of related individuals.

creationism The belief that the world was created by God out of nothing at some time in the recent past and no major changes in organisms have taken place since.

Cro-Magnon Man An early class of humans who lived about twenty-five thousand years ago and whose remains have been found in southern France.

crown The top part of a tooth, or an artificial substitute for this part.

cytology The study of the formation, structure, and function of cells.

D

debridement A medical procedure by which dead tissue is removed from a wound to prevent the development of infection.

deductive reasoning A method of logic that uses specific rules to determine if a statement is true.

denture A partial or complete set of artificial teeth.

direct current (DC) An electrical current that always flows in one direction.

dynamite An explosive discovered by Alfred Nobel that consists of nitroglycerine soaked in diatomaceous earth.

E

elasticity The ability of a material to return to its original size after being stretched.

electric current The movement of electrical charges through some material.

electric shock A medical procedure in which strong electric shocks are administered to a patient's brain in order to change his or her behavior.

electrical generator A device for converting magnetic force to electricity.

electrical motor A device in which electrical current is used to produce some type of motion.

electrochemistry The field of science that involves the relationship between chemical changes and electrical currents.

electrolysis A process by which an electric current is used to bring about a chemical change.

electromagnetic waves Energy waves that consist of two parts, one electrical and one magnetic.

electromagnetism A fundamental physical force of nature that can be observed as both electrical and magnetic phenomena.

electroplating A method by which one metal is laid down on the surface of a second metal.

elephantiasis A disease usually caused by parasitic worms that results in the enlargement of tissues.

energy The capacity to do work; the ability to exert a force over a distance.

enteric fever An infection of the digestive system caused by certain types of bacteria.

epidemic The rapid spread of a disease over a relatively large area so that many people become ill and/or die of the disease in a relatively short period of time.

Esmarch bandage A rubber bandage that can be wrapped tightly around a wounded arm or leg to reduce blood loss.

ether A colorless organic compound that can be used as an anesthetic.

evaporated milk Milk that has been heated so that much of the water it contains is drawn off.

evolution The process by which things change over time.

F

family (biological) A grouping of organisms lower in rank than a class and higher than a genus.

family of elements *See* group of elements.

fermentation Any process by which food is changed when acted upon by yeast, bacteria, or enzymes.

fertilizer Any natural or synthetic material used on crops to supply nutrients missing from the soil or to improve the growth of crops.

food preservation The processes such as drying, freezing, or salting that protect food from decaying.

food-borne disease A disease caused by germs present in food that has spoiled.

force In scientific terms, force is what makes objects change their motion.

fossil Any object, track, imprint, or other remnant of a long-dead organism.

fossil fuel A term used to describe coal, oil, or natural gas.

frequency The rate at which a wave passes a given point in a certain period of time.

G

gene therapy The use of genetic engineering to improve one or more characteristics of an organism.

genes The physical units of heredity. A gene consists of a large group of atoms that carry information about specific hereditary characteristics.

genetic disorder A physical problem, such as a disease, caused by the presence of one or more abnormal genes in an organism.

genetic engineering The process by which the structure of a gene is changed so as to produce some new or desired characteristic.

Words to Know

genetically modified An organism in which one or more genes have been changed.

geometry The branch of mathematics that deals with the properties of figures, such as points, lines, angles, and surfaces.

germ Any microorganism capable of causing disease.

germ theory The scientific concept that infectious diseases are caused by microorganisms, or germs.

greenhouse effect The process by which heat is trapped in Earth's atmosphere by certain chemical compounds, such as carbon dioxide.

group of elements A column of elements in the periodic table with properties similar to each other.

group session A form of treatment in which a small group of individuals meet together, usually with a professional counselor, psychiatrist, or psychologist, to talk about their common mental or emotional problems.

guncotton Another name for the explosive nitrocellulose (or cellulose nitrate), made by reacting nitric and sulfuric acids with cotton.

gunpowder An explosive mixture of potassium nitrate (nitre or saltpeter), charcoal, and sulfur; also known as black powder or blasting powder

H

herbal remedies Naturally occurring plants believed to have beneficial effects on human health.

heredity The process by which certain physical characteristics are passed down from one generation to the next.

hierarchy A group of things organized by rank.

histology The study of tissue of living things.

homeopathy A field of alternative medicine in which patients are treated with herbal remedies that produce symptoms similar to those of their ailment.

humour (or humor) A term once used for fluid that occurs naturally in the body, such as blood or lymph.

hydroelectric plant An energy-producing plant in which running water is used to generate electricity.

I

ice age A period of time in Earth's history during which much of the surface of Earth was covered with ice.

ice pack A large, thick mass of ice.

idealist Someone who believes in the potential for perfection.

immune Resistant to disease.

individual counseling A form of treatment in which a patient talks over his or her problems with a trained professional, such as a psychologist or psychiatrist.

inductive reasoning An argument used in logic that draws on previous experience to make conclusions.

Industrial Revolution Change that came about in western Europe in the middle to late eighteenth century when power-driven machinery was introduced to replace hand labor for the production of many objects and materials.

inert gas A gaseous chemical element that does not react easily with other chemical elements.

infectious Describes illness caused by infection of bacteria, viruses, fungi, or other organisms; contagious.

inheritance of acquired characteristics The theory that the changes that occur in an organism's body can be passed down to the next generation.

intelligence The ability to learn and understand.

internal combustion engine An engine that produces motion when a fuel is burned inside the engine itself.

J

Java Man One of the earliest forms of fossil humans discovered on the island of Java in 1894.

K

kinetic energy The energy of motion.

kinetoscope An early form of a motion picture projector consisting of a wooden box with a peephole in it, through which a person could look to watch a short film loop projected on one end of the box.

L

latex A sappy material produced by rubber trees from which natural rubber is obtained.

law of the conservation of energy Also called the first law of thermodynamics; states that energy can be converted from one form to another, but it can not be created out of nothing or reduced to nothing.

law of triads An early method of classifying elements by groups of three.

Words to Know

leishmaniasis An infectious disease, formerly known as kala azar, caused by protozoans.

linotype A machine in which individual pieces of type are cast when an operator types a letter, number, or symbol into a keyboard.

lip reading A method of communication sometimes used by the hearing impaired in which one learns what a person is saying by watching the speaker's lips.

literacy The ability to read and write.

locomotive A railroad engine that moves under its own power and is used to pull railroad cars.

logic The science that deals with the formal principles of reasoning.

loom A frame or machine used for weaving.

M

maggot therapy The use of maggots to clean out wounds; the maggots devoured dead tissue that might otherwise cause infection.

magnetic field An area of space in which a magnetic force can be detected.

magnetic resonance imaging (MRI) A method for diagnosing disorders of the body by sending radio waves through the body, causing atoms to vibrate in special ways.

malaria A disease very common in tropical countries caused by a parasitic protozoan, causing fevers, chills, and sweating.

meiosis The process by which sperm cells and egg cells divide. Rather than doubling, as cells do in mitosis, sex cells actually split in two. The resulting cells have only half the number of chromosomes as the parent cell.

mental retardation A condition in which a person has an intelligence level significantly below average for someone of his or her age.

meteorologist A scientists who studies weather, climate, and the atmosphere.

meter A device for measuring the flow of electric current or some other electrical property.

microbes Very small living things that can be seen only with the aid of a microscope; they are also called microorganisms or germs.

middle class That group of people in society who are less wealthy than the upper class, but wealthier than the lower class.

mitosis The process of cell division in which the nucleus of the cell duplicates its internal structures and then divides in half so each new cell has the same number of choromosomes as the parent cell.

mobile army hospital A facility for the care and treatment of wounded soldiers that can be set up and moved relatively easily.

moral treatment A philosophy first proposed by Philippe Pinel that recommended a positive social atmosphere for the treatment of mentally ill patients.

N

natural selection The process by which certain organisms are able to survive and reproduce because they are better adapted to an environment than other organisms.

Neanderthal Man A humanlike creature that lived in Europe, northern Africa, and western Asia thirty thousand to one hundred thousand years ago.

neon light A type of lamp that consists of a glass tube filled with neon or some other inert gas that illuminates when electrified.

nitrogen A chemical element that makes up about four-fifths of Earth's atmosphere. It is needed by all growing plants.

nitroglycerine An explosive made by reacting nitric and sulfuric acids with glycerine.

nitrous oxide A gas composed of oxygen and nitrogen that is a mild anesthetic.

Nobel Prizes A group of awards in various fields of study provided for in the will of Swedish businessman Alfred Nobel.

normal curve A bell-shaped graph that represents the probability of occurance of related events.

nucleus An information center in the cell that tells it what to do.

number theory The branch of mathematics that deals with the properties of whole numbers.

O

optical telescope A telescope that can be used for observing astronomical objects by means of the light they give off.

oscillation The regular period motion of an object back and forth between two points.

Words to Know

osteopathy A field of alternative medicine that is based on the belief that if proper balance of life essentials is maintained the body will stay healthy and, ideally, cure itself.

oxidation The process by which a substance reacts with oxygen in the air and changes into a different substance.

P

pericardium The membrane surrounding the heart.

periodic table A chart in which the chemical elements are arranged according to their atomic number and which shows the relationship among the elements.

persistence of vision The tendency of the human eye to "remember" an image for a fraction of a second after the object has actually disappeared.

pessimist Someone who takes a negative view of the things.

pesticide Any natural or synthetic substance used to kill insects and other pests.

phosphorus A chemical element that occurs naturally in many different rocks and minerals. It is needed by all growing plants.

phrenology A medical theory, now known to be false, that the shape of a person's head determined their mental condition.

Pithecanthropus A group of early human-like animals that includes Java Man and Peking Man.

play therapy A form of treatment in which patients are encouraged to express their thoughts and feelings by taking part in some recreational activity.

polonium A radioactive element discovered by Marie and Pierre Curie.

potassium A chemical element that occurs naturally in the Earth's crust. It is needed by all growing plants.

potential energy Energy waiting to be released.

poultice A soft cloth or a preparation of herbs, often warmed, applied to the body as a relief from pain.

prejudice A judgment based on opinions or preconceived notions and not on facts.

professional journal A magazine-like publication that carries articles on specialized topics for people in some particular field of study.

programmed cards Pieces of stiff paper or cardboard in which holes are punched to carry some kind of information, such as the pattern to be woven in a piece of cloth.

Words to Know

protozoan A microscopic, one-celled organism.

psychiatry The field of medicine that deals with mental disorders.

public health The field of medicine that involves health problems within the community that are often the responsibility of some governmental body.

pulp (as for paper) A raw material, made from wood, rags, wastepaper, and other materials, for use in the manufacture of paper.

R

race A group into which humans can be divided based on certain physical characteristics that can be transmitted genetically, such as skin color.

racism The belief that some races are fundamentally different from and superior to other races.

radiation Energy and particles given off by radioactive materials.

radio telescope A telescope that can be used for observing astronomical objects by means of the radio waves they give off.

radio wave A type of electromagnetic wave with a long wavelength and short frequency.

radioactivity The process by which matter breaks apart and gives off radiation.

radium A radioactive element discovered by Marie and Pierre Curie.

respiratory system The body system that includes the organs and tissues involved in breathing, especially the lungs.

rubber A naturally occurring product of the rubber tree, *Hevea brasiliensis*.

S

sanitorium A long-term care hospital, widely used at one time for people with tuberculosis.

sediments Solid matter, such as sand and mud, that settles out of a body of water, such as a lake or river.

seismology The study of earthquakes.

set theory The branch of mathematics that deals with the properties of groups of similar objects and their relationships to each other.

shell of electrons The area in an atom where one or more electrons are found.

shuttle (in weaving) A device for carrying thread from one side of a web to another.

Words to Know

sign language A method of communication often used by hearing and speech impaired individuals that involves the use of standard hand and arm movements to represent words and concepts.

silo An airtight building used for the storage of grains.

smokeless powder A general name given to any explosive that produces relatively small amounts of smoke when ignited, but referring especially to a form of guncotton with this property.

sociology The study of human social behavour.

sonar A method for detecting and locating objects, especially objects under water, by bouncing sound waves off of them.

special needs A term used by educators to describe individuals with special requirements for learning.

species A group of individuals related by descent and able to breed among themselves but not with other organisms.

stable matter Matter that does not break down on its own.

statistics The branch of mathematics devoted to the collection, classification, and interpretation of numerical facts about populations, production, the weather, and other phenomena.

steam engine An engine which is powered by steam.

stick (printer's) A metal form in which individual pieces of type can be arranged to produce a series of words.

stop-action photography A process by which a photograph is taken once every second, or some other unit of time, after which the photographs are combined to produce a running film.

survival of the fittest *See* natural selection.

symbolic logic The branch of mathematics in which logical statements are represented by numbers and symbols.

synthetic chemical Any chemical substance that is produced artificially in the laboratory.

synthetic fertilizer Any compound developed for use in agriculture that provides nutrients to crops.

T

taxonomy The science of classification.

thaumatrope A device consisting of a cardboard disk with images painted on it, in which the images appear to move when the disk is spun.

theory In science, a statement that brings together and summarizes many facts, observations, and laws.

Words to Know

ton-mile A measure equal to a ton of goods carried a distance of one mile.

toxic Poisonous.

tracer A material whose presence can be detected because of the radiation it gives off.

trephination A very old medical practice in which certain disorders of the brain are treated by making holes in the patient's skull.

triage The process by which doctors decide who among a group of wounded and injured individuals is most in need of medical care.

tropical medicine The field of medicine focused on diseases and disorders that occur primarily, but not exclusively, in hot, moist parts of the world.

tuberculosis (TB) An infectious disease transmitted by bacteria that affects the human respiratory (breathing) system.

type (as in printing) Metal forms that contain the letters, numbers, and symbols from which printed pages are produced.

U

ultrasound Vibrations with properties similar to those of sound, but with frequencies beyond the range of human hearing.

uniformitarianism A geological theory that says that Earth as we know it today has been produced by means of processes that still occur and can still be observed today.

V

vaccine A preparation containing dead or weakened germs that is injected into a person's blood stream to protect against a disease by building immunity to those germs.

"vis viva" A seventeenth-century scientific term to describe an object's potential for motion, and its ability to exert force on another object.

vitamin-deficiency disease A disease that develops when a person does not receive an adequate supply of a vitamin.

vulcanization The process by which sulfur is added to liquid rubber to make the rubber stronger and more resilient.

W

wavelength The distance between two peaks or two valleys in a wave.

wireless telegraph A device for sending messages by means of radio waves transmitted between two stations.

work The ability to exert a force over a distance.

Words to Know

working fluid The liquid or gas used in a refrigeration system to carry heat away from the system.

X

X rays A very powerful form of electromagnetic radiation with very short wavelengths and very high frequencies.

Y

yellow fever A highly infectious disease caused by a virus transmitted by mosquitoes.

Z

zoetrope A device containing a strip of images affixed around the outside edge of a rotating drum; the images appear to move when the drum is rotated.

Science, Technology, and Society

The Impact of Science in the 19th Century

chapter three Mathematics

Chronology **164**
Overview **165**
Essays **168**
Biographies **182**
Brief Biographies **200**
Research and Activity Ideas **207**
For More Information **208**

163

CHRONOLOGY

1801 German mathematician Johann Karl Friedrich Gauss (1777–1855) publishes *Disqusitiones Arithmeticae* (Arithmetical Researches), which greatly expands number theory.

1810 The first privately established mathematics journal is founded at the École Polytechnique in France by Joseph Diaz Gergonne (1771–1859). It is called the *Annals of Pure and Applied Mathematics*.

1812 French astronomer and mathematician Pierre-Simon Laplace (1749–1827) publishes *Théorie Analytique des Probabilités,* and establishes the basis for the modern science of probability.

1822 French mathematician Jean-Baptiste Joseph Fourier (1768–1830) discovers a method for analyzing periodic motion. Eventually known as Fourier series, it proves invaluable to the study of all types of waves.

1826 German mathematician August Leopold Crelle (1780–1855) launches the *Journal for Pure and Applied Mathematics*. Known as *Crelle's Journal*, it is the first periodical to focus exclusively on research in pure mathematics.

1828 English mathematician George Green (1793–1841) publishes his *Essay on the Application of Mathematical Analysis to the Theories of Electricity and Magnetism*. It eventually provides the basis for understanding electric and magnetic fields.

1847 English mathematician George Boole (1815–1864) writes the first paper to suggest that logical ideas can be expressed using mathematical symbols; seven years later he fully develops his theories, which are now known as Boolean algebra.

1854 German mathematician Georg Friedrich Bernhard Riemann (1826–1866) proves that various types of non-Euclidian geometry are possible.

1865 German mathematician August Ferdinand Möbius (1790–1868) invents the Möbius strip, a figure that has one side and one edge. The discovery contributes to the establishment of a new branch of mathematics known as topology (the mathematical study of surfaces).

1865 The London Mathematical Society is formed.

1874 German mathematician Georg Ferdinand Ludwig Philipp Cantor (1845–1928) publishes his first paper on set theory and the theory of the infinite.

1888 The first professional mathematics society in the United States is created. Originally called the New York Mathematical Society, it becomes the American Mathematical Society in 1894.

OVERVIEW

Chapter Three

MATHEMATICS

By the end of the eighteenth century, although advances had been made in a number of mathematical fields, including analytic geometry and algebra, mathematical thought was dominated by the calculus. Discovered in the late seventeenth century by English physicist and mathematician Isaac Newton (1642–1727) and German mathematician Gottfried Leibniz (1646–1716), mathematicians were fascinated by the power of the new tool they had been given.

Many of the best mathematicians of the day explored the way calculus could be used to solve physical problems in the real world. They were excited by the prospect of using numbers to explore such indescribable concepts as space and time. Most, however, struggled to answer a fundamental question: although calculus worked, how could it be proven? As scientists attempted to answer that question, they began to challenge the very nature of mathematics.

Mathematicians of the nineteenth century built upon eighteenth century advances and made such incredible new discoveries that the period is called the Golden Age of mathematics. According to historians, the advances made during those one hundred years outweigh all the mathematical discoveries previously made. The nineteenth century is often compared to the time of the great Greek mathematician Euclid (c. 330–c. 260 B.C.), during which the first great theories of mathematics were developed.

New fields of mathematics discovered

During the nineteenth century, completely new fields of mathematics were discovered. An example of a new field of mathematics is non-Euclidean geometry. For centuries, mathematicians had believed there could be only one kind of geometry. The laws of that geometry had been established by Euclid in the third century B.C. During the 1800s, mathematicians such as Janos Bolyai (1802–1860), Nicolai Lobachevsky (1793–1856), and Carl Friedrich Johann Gauss (1777–1855), challenged Euclid's laws and proved that there could be other forms of geometry.

By doing so, Lobachevsky, Bolyai, and Gauss, radically changed the way mathematics was viewed. Euclid's laws had been accepted without question for thousands of years. If mathematicians were able to prove that

Mathematics

OVERVIEW

there could be new kinds of geometry, why not new kinds of algebra or new kinds of trigonometry?

German mathematician Georg Cantor (1845–1918) was another individual who challenged one of the basic beliefs in mathematics. For centuries, scholars had considered the idea of infinity to be just that, an idea. Infinity was a limitless concept that could not be handled with numbers. Cantor, however, developed the theory of sets, which created a mathematical framework for dealing with problems of space and time. Cantor amazed his colleagues with the notion that there could be more than one kind of infinity and that some infinite sets are larger than others.

Other nineteenth century mathematicians worked on subjects such as hypercomplex numbers, projective geometry, and elliptical functions. These complicated fields were created by men and women who were fascinated with operations that could be carried out by numbers and letters. The amazing fact is that these explorations in pure mathematics often led to totally unsuspected practical applications. For example, mathematicians had long been intrigued by the idea of imaginary numbers, numbers that contain the square root of -1 (what number when multiplied by itself would give you -1). No one could quite imagine what applications, if any, such numbers might have. Yet, a century later they were being used to solve problems in many fields of science, including electrical engineering and wave mechanics.

Mathematical discoveries with practical uses

Other mathematical discoveries eventually had very practical uses, as well. Some helped revolutionize modern society. One example is Boolean algebra, developed by the English mathematician George Boole (1815–1864). Boole proved that logical statements could be represented by mathematical symbols. A century later, this idea became one of the fundamental tools used to develop computer science.

English mathematician George Green (1793–1841) developed a mathematical theory to explain electric and magnetic fields. His theory helped future scientists create practically every modern-day electrical device. French mathematician Jean-Baptiste Joseph Fourier (1768–1830) discovered a method for analyzing complex periodic (repeating) motion, such as the motion of waves. Known today as Fourier series, this method is used today to analyze all types of wave phenomena, including sound waves and light waves.

Mathematicians delve deeper into known fields

During the nineteenth century, mathematicians not only developed new fields of mathematics, they expanded on existing mathematical ideas that had not been fully developed. For example, calculus, the area of mathematics that prevailed during the eighteenth and nineteenth cen-

Mathematics

OVERVIEW

turies, was finally given a serious foundation by mathematicians, including Augustin Louis Cauchy (1789–1857), Georg Riemann (1826–1866), and Karl Weierstrass (1815–1897).

Mathematicians had always been intrigued by the properties of numbers, a field known as number theory. That field saw striking advances during the 1800s thanks to the work of Swiss mathematician Leonhard Euler (1707–1783) and German mathematicians Gauss, Riemann, Peter Dirichlet (1805–1859), and Ernst Kummer (1810–1893).

Probability theory was another subject that had been investigated by mathematicians for many decades. Prior to the nineteenth century, it focused primarily on problems having to do with games of chance. During the 1800s, mathematicians and physicists were able to use probability to solve problems in physical sciences, such as astronomy and physics. For example, French mathematician Pierre-Simon Laplace (1749–1827) used the theory of probability to predict planetary orbits. Others, such as Lambert Adolphe Quintelet (1796–1874) used statistical data to further the social sciences.

Mathematics becomes a profession

In the midst of the incredible changes taking place during the nineteenth century, mathematics became recognized as a true field of study. Prior to 1800, there were no professional mathematicians and mathematical research was conducted by individuals outside the university setting. Mathematical discoveries were acknowledged only when they could be practically used in some other field. By the end of the century, a branch of pure mathematics had evolved and mathematical research was accepted as a legitimate endeavor apart from its practical application.

Universities began to develop a mathematics curriculum, which made it necessary for the creation of mathematics departments. And research was brought into the university setting with mathematicians (always men at first) hired for the sole purpose of conducting research. At the same time, the scope of mathematics education was expanded. As technology developed, advanced study in mathematics became essential for anyone who was interested in becoming a physicist, astronomer, chemist, or any other kind of physical scientist.

As mathematics grew into a true profession, mathematicians began to organize into a mathematical community. Before the nineteenth century, most mathematicians worked independently on their research. They communicated the results of their work by writing letters to each other. They seldom met in large groups to discuss the advancements that were taking place, and there were no formal university programs to train an individual to become a mathematician.

Mathematics
ESSAYS

As the nineteenth century progressed, all that changed. The first professional mathematical societies were created in France, which before the 1800s, had been the center of mathematical activity. By the mid-1800s, however, mathematical societies were formed in England and Germany. In 1888, the New York Mathematical Society was created, which was the first professional mathematical society in the United States. Through regular meetings, members who belonged to these societies could share their findings and compare their ideas.

These societies also published the first journals devoted exclusively to mathematics. In 1810, the very first privately produced mathematical journal was established in France. Called the *Annals of Pure and Applied Mathematics*, it was created by French mathematician Joseph Diaz Gergonne (1771–1859). These journals helped keep mathematicians informed about new advancements.

Throughout the nineteenth century, mathematics was transformed. The century began with mathematicians eager to break new ground thanks to the invention of the calculus. It ended with discoveries that helped mathematics evolve into a mature profession. Mathematicians, such as Gauss, Fourier, Cauchy, and Laplace, pushed mathematics into many new directions and changed the way individuals thought about numbers, figures, and the physical world.

ESSAYS

EARLY ATTEMPTS TO FORMULATE A MATHEMATICAL THEORY OF ELECTRICITY

Overview
Prior to 1800, scientists had conducted many experiments involving electricity and magnetism. They did not, however, fully understand how the two forces functioned, and no one had discovered a way to explain electricity and magnetism in mathematical terms. That situation changed completely in 1828 when amateur mathematician George Green (1793–1841) published his groundbreaking essay on electric and magnetic fields.

Background
The forces of electricity and magnetism had been investigated by humans for thousands of years. The ancient Greeks, for example, theorized about the nature of static electricity (electrical charges on a material) and natural magnets found in the Earth.

Mathematics

ESSAYS

By the nineteenth century, interest in these two basic forces increased considerably. A number of important experiments produced new information about electricity and magnetism and demonstrated the connection between them. In 1800, Italian physicist Count Alessandro Volta (1745–1827) introduced the first electric battery to the world. Known as the Voltaic pile, it consisted of zinc and silver disks separated by felt soaked in brine. It was considered the first source of a continuous electric current.

In 1820, Danish physicist Hans Christian Oersted (1777–1851) showed that a wire carrying an electric current will generate a magnetic field around itself. A year later, English physicist and chemist **Michael Faraday** (1791–1867; see biography in Physical Sciences chapter) demonstrated the reverse process in which a magnetic field can be used to generate an electric current.

These results prompted scientists to think about two basic questions. First, how could an electric or magnetic field be represented mathematically? A century earlier, English physicist and mathematician Isaac Newton (1642–1727) had demonstrated that gravitational fields could be represented by mathematical equations. The logical assumption was that electric and magnetic fields could also be described by equations.

The second question was, what is the connection between electric and magnetic fields? It seemed clear that the answer to this question could be found if the first question was solved. That is, a mathematical theory of electric fields would also explain how magnetic fields are associated with them.

The first successful attempt at answering these two questions was taken on by George Green, an English miller with a strong interest in mathematics, but no formal training. Green's research resulted in an essay titled, "Essay on the Application of Mathematical Analysis to the Theories of Electricity and Magnetism." His work was financed by fifty-two friends and patrons who received copies of Green's essay after it was completed in 1828. It was common during the time for scientific researchers to be financed by wealthy individuals who were interested in scientific advancement.

Because Green's essay was not widely published, it probably would have been forgotten. In 1850, however, nine years after Green's death, it was discovered by a very influential individual. That individual was William Thomson, Lord Kelvin (1824–1907), who was one of the leading physicists of his day. While conducting his own research, Kelvin came across a reference to Green's essay. Recognizing its importance, Kelvin arranged to have it published in a German journal of mathematics. As a result, Green's work had a remarkable effect on researchers of his day.

In his work, Green developed a set of mathematical equations that describe electric and magnetic fields. The set of equations is now known

George Green (1793–1841)

During his lifetime, English mathematician George Green developed some of the most important theories in the history of mathematical physics. He is responsible for publishing one of the most influential mathematical papers of the nineteenth century, which provided the basis for understanding electric and magnetic fields. The amazing thing was that Green had no formal schooling, had little time to spend on his research, and was a self-taught mathematician.

Green was born in 1793 in Nottingham, England, the son of a prosperous baker. His formal education consisted of one year at a local Nottingham school when he was eight years old. Some historians believe Green was then tutored in mathematics and French by John Toplis, a student attending nearby Cambridge. It was especially important that Green was instructed in French since at the time, the greatest advancements in mathematics were taking place in France and all the major papers of the day were written in French.

Green worked full-time in his father's mill and eventually took over the business. He had seven children and little free time to devote to mathematics. Green, however, continued to study on his own.

In 1828, Green published what would become the most important work of his life, *Essay on the Application of Mathematical Analysis to the Theories of Electricity and Magnetism*. Sir Edward Bromhead, one of the subscribers

as Green's theorem. Green also suggested for the first time the concept of potential energy, the capacity of an object to perform work. That concept is a fundamental aspect of physical science today.

Impact

Green's research was only the first step in creating mathematical theories of electricity and magnetism. These efforts continued through the 1800s and culminated in the work of the Scottish physicist **James Clerk Maxwell** (1831–1879; see biography in Physical Sciences chapter). In 1864, Maxwell applied the mathematical techniques of Green to the experimental observations of Faraday. In doing so, he developed a set of four equa-

who purchased the essay, was amazed at the level of Green's work and soon began a correspondence with the self-taught mathematician. Because of Bromhead's influence, Green published three major papers between 1830 and 1834. None of them received much attention.

During this time, Green had very little confidence in his abilities and doubted very much that his essays would ever have any impact. In 1833, when he was forty years old, Green decided to become a student at Caius College of Cambridge University. There he concentrated in mathematics and earned his degree in 1837.

After graduation, Green published two more papers and received a fellowship to teach at Cambridge. Unfortunately, he fell ill in 1840 and died one year later at the age of forty-eight.

A few years after his death, Green was given the recognition he was denied during his lifetime. When conducting his own research, physicist William Thomson (Lord Kelvin; 1824–1907), came across Green's 1828 essay. He arranged to have it published in an influential journal and shared the essay with many of his contemporary scientists and mathematicians. It was in this way that the work of a little-known, self-taught man impacted the future of mathematics.

tions that fully explain light in terms of electric and magnetic fields. According to Maxwell, electric and magnetic fields are bound together. When a magnet moves, it produces an electric current; when an electric current flows, it causes a magnetic force. As electricity and magnetism interact, they create waves in space. Maxwell identified these waves as light waves. (See essay "Nineteenth Century Advances in Electromagnetic Theory" in Physical Science chapter.)

The impact of Green's (and Maxwell's) work was enormous. Their theories contributed to the development of electric motors and generators, which led to rapid advancements in industrialization. In 1888, Maxwell's equations helped Heinrich Hertz (1857–1894) produce and detect radio waves, which eventually launched the possibility of mass communication. (See essay "Heinrich Hertz Discovers Radio Waves" in the Physical Science chapter.) In the 1900s, the equations set the stage for Albert Einstein's (1879–1955) theory of relativity.

In the modern world, almost every electrical device makes use of the mathematical equations developed during the nineteenth century. Everything from the giant dynamos that generate electricity to pocket-size portable radios operate on the electromagnetic theory. Other forms of electromagnetic waves include X rays, ultraviolet light, infrared radiation, and radio and radar waves. Every time you turn on a television, talk on a cell phone, or microwave your lunch, you are able to do so because of Green and Maxwell.

■ DEVELOPMENTS IN MATHEMATICS EDUCATION

Overview

Because of economic and political changes in the Western world, the nineteenth century witnessed a dramatic revolution in the teaching of mathematics. At the beginning of the century, mathematics was traditionally taught to upper class men at the university level. By the end of the century, the need for technical workers and military officers capable of understanding new technologies sparked the creation of technical institutions and mathematical societies.

Background

Instruction in mathematics has always been an essential part of a well-rounded education. Even 2,400 years ago, the Greek philosopher Plato (c. 428–c. 348 B.C.) refused to admit students to his academy unless they were trained in mathematics. By the 1400s, three areas of mathematics gradually became part of the standard liberal arts education expected of European upper class men (not women). These three areas were arithmetic, geometry, and logic.

Mathematics texts were among the first books printed with moveable type, which was invented in 1454. For example, the earliest arithmetic text was published in Trevisi, Italy, in 1478. Four years later, the first printed version of Euclid's geometry appeared in Venice. These texts were used in schools, colleges, and universities.

Prior to the nineteenth century, however, mathematics taught in schools and colleges was generally not designed to help students in their future careers. An understanding of mathematics was simply part of a young gentlemen's training. They were also expected to master subjects, such as history, Latin, and Greek. At the time most wealthy young men were destined to serve in the military as officers or become members of the clergy.

Words to Know

arithmetic: The branch of mathematics that involves calculations, such as addition, subtraction, multiplication, and division, with real numbers.

geometry: The branch of mathematics that deals with the properties of figures, such as points, lines, angles, and surfaces.

logic: The science that deals with the formal principles of reasoning.

number theory: The branch of mathematics that deals with the properties of whole numbers.

professional journal: A magazine-like publication that carries articles on specialized topics for people in some particular field of study.

set theory: The branch of mathematics that deals with the properties of groups of similar objects and their relationships to each other.

Professional mathematicians had virtually no place in the academic world and mathematical research was conducted outside the university setting. Researchers were usually supported by wealthy patrons or by professional scientific and mathematical societies. They studied mathematics largely because they were fascinated by numbers and symbols; they did not expect their work to have much impact on the daily lives of ordinary people.

The Industrial Revolution, which began in Great Britain in the mid-1700s, completely altered the nature of mathematics. As new technologies changed the world of business and industry, it was necessary for workers and managers at nearly every level to deal with numbers on a daily basis. Mathematics was called upon to figure weights and measure and to calculate currency exchanges.

As trade expanded, it was essential to develop advanced methods for goods to be transported. Engineers had to tackle problems such as how much stress could be placed on a railroad bridge or the maximum amount of pressure a boiler on a steamship could withstand. Mathematics was no longer a luxury that upper class men pursued or that scholars speculated

Mathematics
ESSAYS

about within the walls of academia. It had become an essential tool of the everyday workplace.

Impact

The first nation to understand how important mathematics would be in a newly industrialized world was France. In 1794, the national government authorized the creation of technical schools known as the École Polytechnique, which would produce most of the important members of the French scientific community. At the same time, the École Normal Superiore was established to train teachers. Among the illustrious members of the staff was mathematician and astronomer **Pierre-Simon Laplace** (1749–1827; see biography in this chapter).

Change in the United States took place much more slowly. During the late 1700s, no U.S. college required extensive study of mathematics. Geometry was not a required course until the Civil War (1861–65). And it was not until the last quarter of the nineteenth century that schools and colleges began to offer mathematical instruction that was practical in nature. In 1872, Harvard, Yale, and Dartmouth colleges adopted a new text titled, *A New and Complete System of Arithmetic Composed for the Use of Citizens of the United States*. It included information on commercial mathematics, currency conversions, weights and measures, the calendar, and interest calculations.

By the end of the 1800s, state universities and technical schools were established that offered students the opportunity to study new fields of engineering and helped contribute to a growing new U.S. mathematical community.

Outside the university setting, mathematics was emerging as a true profession. Mathematicians began forming professional societies and publishing professional journals. Again, the French led the way. The first privately established journal devoted exclusively to mathematics was founded at the École Polytechnique in 1810. It was called the *Annals of Pure and Applied Mathematics*. In 1826, the first German language mathematics journal titled, the *Journal for Pure and Applied Mathematics,* was established by August Leopold Crelle (1780–1855). Both journals still exist today.

In 1841, the Berlin Mathematical Society began publishing *The Archives of Mathematics and Physics with Special Consideration of the Needs of Teachers*. It was the first journal devoted to the teaching of mathematics. In 1865, the London Mathematical Society was formed.

The first professional mathematics society in the United States was created in 1888. Originally called the New York Mathematical Society, it is now known as the American Mathematical Society.

Mathematics

ESSAYS

In 1875, the National Teacher's Association (now the National Education Association) was established to address the needs of teachers at the primary and secondary school levels. Between 1892 and 1912, special committees were formed to make recommendations about mathematics education. In 1920, a specialized organization, the National Council of Teachers of Mathematics, was created to focus exclusively on mathematics education.

The strongest reformer of mathematics at the elementary school level was the Swiss teacher Johann Heinrich Pestalozzi (1746– 1827). Pestalozzi outlined a method of teaching mathematics to younger children. He wrote that children should begin their mathematical education with topics that are familiar and concrete. Only when they are comfortable with such basic concepts as addition and multiplication should they finally move on to abstract, higher concepts of mathematics. Pestalozzi's philosophy had a strong influence on mathematics education in his native Switzerland and, later, in the United States.

The strongest reformer of mathematics at the elementary school level was the Swiss teacher Johann Heinrich Pestalozzi. Pestalozzi outlined a method of teaching mathematics to younger children. (Reproduced by permission of Archive Photos, Inc.)

Just as economic and social factors affected the development of mathematics in the nineteenth century, the debate over mathematics education continues today. Some educators argue that schools should emphasize applied mathematics, such as consumer arithmetic, computer skills, and other forms of mathematics that people use in their everyday lives. They insist that as technology continues to advance at such a rapid speed, a practical math background is necessary for students to compete in the job market. Other educators believe that it is still essential for students to learn fundamental principles of abstract mathematics, such as set theory.

■ ADVANCES IN THE MATHEMATICS OF LOGIC

Overview

Logic is a system of reasoning used to test the truth of various statements. It was originally considered to be a branch of philosophy, but in the nineteenth century, mathematicians argued that logic could be applied to mathematics, as well. By using mathematical symbols to study problems in logic, they created the foundations for the digital processes that drive most of today's technology.

Mathematics

ESSAYS

Background

There are two types of logical arguments: *inductive* reasoning and *deductive* reasoning. Inductive reasoning draws upon previous experience to reach probable conclusions. For example, a person might reason as follows: (1) The present month is January. (2) The weather is usually cold and snowy in January. (3) Therefore, this month will probably be cold and snowy.

Deductive logic does not rely on experience to reach conclusions that are *probably* true. Instead, deductive logic uses specific rules to determine whether a statement is *actually* true. Specifically, it begins with statements or premises that lead to a conclusion. All the premises in a deductive argument must be true in order for the final conclusion to be valid. For example, a person may argue that: (1) All mammals are blue. (2) A cow is a mammal. (3) Therefore, all cows are blue. The final statement is false because one of the two premises is not true: a cow may be a mammal, but all mammals are not blue.

The first person to create a system of rules for logical reasoning was the Greek philosopher Aristotle (384–322 B.C.). He collected these rules in a set of books called the *Organon,* or "Instrument." According to Aristotle, all arguments can be reduced to a sequence of three propositions: two premises and one conclusion. A set of three statements, such as the blue cow example, is known as a *syllogism.* Over the centuries, the study of logic established by Aristotle became an essential part of formal education.

By the seventeenth century, a few scholars believed that logic be used in other fields besides philosophy. German philosopher and mathematician Gottfried Wilhelm Leibniz (1646–1716) was one of these scholars. Leibniz suggested that simple statements could be expressed using numbers and symbols. He argued that logical arguments could, therefore, be represented as mathematical problems.

Leibniz worked on his "calculus of reasoning" theory for over thirty-five years, but it attracted very little attention. It would be more than 150 years before another mathematician picked up where Leibniz left off.

In the 1840s, English mathematician **George Boole** (1815–1865; see biography in this chapter) began investigating how algebra and logic were connected. He published his findings in an 1847 pamphlet titled, *Mathematical Analysis of Logic.* In 1854, Boole outlined his system in the book, *An Investigation into the Laws of Thought.* He explained what Leibniz had originally attempted to prove: that the laws of reasoning could be expressed as mathematical symbols.

Boole developed what eventually became known as symbolic logic. His system also formed the basis for an entirely new type of algebra, known as

Words to Know

binary system: A method of counting that uses only two digits, 1 and 0.

deductive reasoning: A method of logic that uses specific rules to determine if a statement is true.

inductive reasoning: An argument used in logic that draws on previous experience to make conclusions.

symbolic logic: The branch of mathematics in which logical statements are represented by numbers and symbols.

Boolean algebra, or the algebra of sets. Boolean algebra is a binary system, which means it is a system based on two values. Boole's two values included a universal, or positive, set and an empty, or negative, set. The positive and negative values are represented in a number of ways, such as 1 or 0, true or false, or on and off. Boole used symbols in algebra, such as x, y, and z, to stand for subsets of the universal positive or negative sets. Mathematical operational symbols, such as the addition sign (+) or the multiplication sign (x) represented the relationship between the statements or ideas.

The development of symbolic logic was taken even further by other mathematicians of the nineteenth century. German mathematician and philosopher Gottlob Frege (1848–1925) attempted to update Leibniz's "calculus of reasoning" in his 1879 booklet, *Begriffsschrift*. Considered to be the founder of modern mathematical logic, he developed an entirely new set of symbols to link mathematical statements in logic.

Frege's work was not well received, partly because the symbols he used were so complicated and confusing. Another mathematician who had greater success was the Italian Giuseppe Peano (1858–1932). Peano introduced symbols that were similar to the ones suggested by Frege. Peano's symbols, however, were much simpler and easier to understand.

Peano convinced a number of colleagues to experiment with his system of symbolic logic. Eventually, he and other mathematicians produced a five-volume publication, *Formulary of Mathematics*, which summarized their work. The publication would serve as a foundation for all areas of mathematics.

Boolean Logic Every Day

Early in the twentieth century, Boolean algebra was used primarily by mathematicians, engineers, and computer programmers. By the end of the century, personal computers became more available, and most people started to use Boolean logic almost every day—in the form of Boolean searching.

When searching for information in databases or on the Internet, Boolean instructions, known as operations, remain the most popular method for achieving the best search results. The most well-known Boolean operations are AND, OR, and NOT. When applying any of these operations to a search, it means you are asking whether or not objects share specific properties.

Suppose that you are searching a used-car database on the Internet for information about sports cars. By using the operation AND in your search, as in color=red AND model=Corvette, you are asking for information only about Corvettes that are red.

If the operation OR is used in the model type, such as color=red and model=Corvette OR model=Mustang, you will receive information either about red Corvettes or red Mustangs.

If you want information about all models of sports cars except for Corvettes, you would use the NOT operation, as in model=NOT Corvette.

Impact

The creation of symbolic logic was significant because it applied mathematical laws to philosophical reasoning. By replacing ordinary language with mathematical symbols, mathematicians hoped to create a universal language that did not leave room for misinterpretation. They also believed that creating mathematical rules for logic would transform the philosophy of logic into a science.

During the nineteenth century, symbolic logic had a relatively modest impact on the daily lives of men and women. By the mid 1900s, however, one form of symbolic logic, Boolean logic, had a profound effect on society.

In the 1930s, scientists discovered that Boole's binary system could be easily applied to many physical systems. An electronic circuit, which is composed of millions of tiny electric switches, is an example of such a system. Each electric switch can be turned "on" or "off" to indicate whether or not a current is passing through a circuit. The position of the switch could be represented numerically as "1" for the "on" position and a "0" for the "off" position.

Boolean logic was eventually used to create digital circuits of all kinds, which are used to power objects such as coffeepots, cameras, stereo systems, and, most importantly, computers. Computers operate on binary systems in which circuits are either "on" (represented as "1") or "off" (represented as "0"). Boolean algebra is used to convert language into numbers, which can then be processed by computers. This means that every time you type something into your keyboard, the computer reads a series of 1's and 0's and circuits are turned on or off.

During the nineteenth century, Leibniz, Frege, Boole, Peano, and their colleagues began their studies of symbolic logic because they were intrigued by numbers. It is amazing, then, that their legacy helped launch the computer age of the late-twentieth century.

■ THE RISE OF PROBABILITY THEORY

Overview
The development of the probability theory in the nineteenth century had a profound effect not only on mathematics, but on a number of fields, including insurance, economics, medicine, and physics. Probability theory is a mathematical tool that allows scientists to analyze an event and predict its outcome. For example, the laws of probability can be used to predict how frequently a tossed coin will turn up heads.

Background
The beginning of mathematical probability can be traced to a series of letters exchanged between two French mathematicians, Pierre de Fermat (1601–1665) and Blaise Pascal (1623–1662). Their famous correspondence centered on a very practical everyday problem: how to divide money among players who had wagered on a game of chance if the game was interrupted before it was completed. By analyzing games of chance and attempting to predict the odds of winning, the two men created mathematical techniques for predicting probability.

The correspondence between Fermat and Pascal came to the attention of Dutch physicist Christiaan Huygens (1629–1695). In 1657, Huygens

Words to Know

normal curve: A graph that shows the distribution of one variable compared to a second variable.

statistics: The branch of mathematics devoted to the collection, classifying, and interpreting of numerical facts about populations, production, the weather, and other phenomena.

published the first text on probability. In 1713, a book titled, *The Art of Conjecturing,* written by Swiss mathematician Jakob Bernoulli (1654–1705) was released.

The first book on probability written in English was *The Doctrine of Chances,* written by French mathematician Abraham De Moivre (1667–1754) in 1718. In his book, De Moivre introduced the concept of a normal curve, now known as the bell curve. A normal curve is shaped like a bell, with a peak in the middle and rapidly sloping curves on either side. Its shape can be used to graphically represent many events studied in probability. For example, if a two-sided coin is flipped many times, the chances that it will come up heads is equal to the chances of it coming up tails. That probability is the highest and forms the central peak of the normal curve. The probability of the coin always coming up heads (or always coming up tails) is very low and forms the outer lower edges of the normal curve.

Prior to the 1800s, most books, including the ones written by Huygens, Bernoulli, and De Moivre focused on gambling problems and used gambling examples and terms. But, just as the shape of other sciences was changing, so, too, was the field of probability. The changes were due, in large part, to the work of English physicist and mathematician Isaac Newton (1642–1727). Newton had created a new scientific approach that relied on careful experimentation and explanation of events based on quantitative, or measurable, results. In effect, he took the guesswork out of science.

The need to observe, measure, and explain was especially important in the growing field of probability. It also helped to transform it into a legitimate science. Prior to the nineteenth century, people felt that probability did not belong in the sciences. Since it was a science that produced results that were only probably true, it was believed to be imperfect.

Mathematicians, such as **Pierre-Simon Laplace** (1749–1827; see biography in this chapter), however, pointed out that there were some instances where absolute knowledge or absolute certainty was impossible. Laplace believed that in those instances, applying the rules of probability allowed for a reliable degree of certainty. Like Newton, nineteenth-century scientists began using carefully analyzed statistical data to predict probable outcomes.

Mathematics

ESSAYS

Impact

As the laws of probability were accepted, nineteenth-century mathematicians and scientists began to see some practical applications for probability theory. For example, it became an essential tool of life insurance companies. Statistics, such as gender (male or female) and age, were gathered and used to create mortality tables. The tables were used to predict the average life expectancy of someone purchasing a life insurance policy. The table also helped to set rates for the policies issued.

Analyzing social statistics was greatly advanced by the Belgian astronomer Lambert Adolphe Quetelet (1796–1874), who was convinced that many events in everyday life could be studied with probability theory. He analyzed all kinds of statistics about humans and social groups, such as the height and weight of individuals and the number of crimes that occur in a society. He concluded that many of these statistics fit De Moivre's normal curve.

By placing human statistics on a bell curve, Quetelet believed he could accurately describe the average man. For instance, he placed people of average height at the peak of the bell curve, while smaller numbers of very tall people were placed on one side of the curve and very short people were placed on the other side of the curve. Quetelet referred to his studies as social physics. He considered social physics to be a true science and he borrowed terms from statistical astronomy to reinforce the similarities. For instance, any individual who deviated from the "average man" model was considered to be an "error."

Quetelet was interested in creating stable and predictable data out of the chaos of society. Given a large enough body of data, such as a large group of people, Quetelet was positive that he could always come up with

Pierre de Fermat, a French mathematician who helped establish mathematical techniques for predicting probability. (Reproduced by permission of Corbis-Bettmann.)

Mathematics

BIOGRAPHIES

a constant average. This theory is known as the "law of large numbers." By using the law of large numbers, Quetelet believed data (such as data on crime and disease) could be gathered and used to help improve society.

The approach developed by Quetelet was adopted by social scientists into the twentieth century. It also greatly influenced scientists in a number of other fields. For instance, English scientist Francis Galton (1822–1911) used Quetelet's theories to advance the study of heredity, which is the science of how and why characteristics are passed from parent to offspring.

Galton also developed a new method for studying statistics called *correlation*. Correlation is used to measure how one thing changes in relation to another. For example, scientists might measure the correlation between the increase in the number of smokers in a population and the increase in the number of individuals suffering from lung cancer.

Prior to the nineteenth century, probability was used primarily to predict gambling outcomes. By the end of the nineteenth century, the probability theory had become a powerful tool for evaluating data. It was used to revolutionize a number of scientific fields, including physics, chemistry, astronomy, and medicine. It could also be applied to everyday situations.

More importantly, probability changed the very nature of science. Before the 1800s, scientists believed that their work should focus only on the search for absolute truth. By the end of the century, they realized that some things could only be known with a certain degree of scientifically measured probability.

Adolphe Quetelet applied the laws of probability to humans and social groups. (Courtesy of the Library of Congress.)

BIOGRAPHIES

CHARLES BABBAGE (1791–1871)

Charles Babbage was an English mathematician and inventor who designed and built some of the first machines for doing mathematical calculations. He is considered the inventor of the first mechanical computer.

Babbage was born the son of a banker in London, England, on December 26, 1791. At a young age, he became especially interested in studying algebra. He was so interested in mathematics, that he studied it even in his free time.

In 1810, Babbage enrolled at Cambridge University and majored in mathematics. He was dismayed by the lack of quality mathematics education in Great Britain and helped found the Analytical Society in 1812. Its goal was to introduce the best of European mathematical ideas to England. In 1815, when he was twenty-four years old, he was elected to the Royal Society, an organization that included the best scientists in England.

It is believed that Babbage first thought of calculating machines during a meeting of the Analytical Society in 1820. According to legend, he was studying a logarithm table (a table that helps to multiply and divide large numbers) that contained a number of errors and said, "I am thinking that all these mathematical tables might be calculated by machinery."

Babbage was determined to create a cheaper, faster, and more accurate way of producing mathematical tables. By 1822, he had built a small hand-cranked machine that could perform mathematical calculations. His invention was so impressive that he received funding from the British government to produce a full-scale machine that would calculate navigational tables for the British Navy. It was called a difference engine, and work on it proceeded very slowly. Since the technology did not exist, Babbage had to invent all of the mechanical functions needed to operate it. By 1834, the difference engine was still not complete.

Babbage, however, remained optimistic and began developing an idea for an even more complex calculating maching—the analytical engine. While his difference engine was expected to perform only one function at a time, the analytical engine could perform many functions at once. It was designed to be steam-driven and used punch cards that instructed the machine to perform mathematical functions. On paper, Babbage had created the very first computer.

In 1871, Babbage died before a real analytical engine was ever built. The British government had refused to finance another of his projects and

Mathematics

BIOGRAPHIES

Charles Babbage. (Courtesy of the Library of Congress.)

Mathematics
BIOGRAPHIES

Charles Babbage's "Difference Engine." (Reproduced by permission of The Granger Collection.)

his own money had run out. In addition, the technology needed to build such a machine simply did not exist in the nineteenth century. Babbage was, indeed, a man born ahead of his time.

During his lifetime, Babbage did make many important contributions. For example, he helped develop the modern postal system and he devised a signaling code for lighthouses. In addition to the Analytical Society, he helped create a number of scientific societies, including the Royal Astronomical Society (1820) and the Statistical Society of London (1834).

In 1828, his reputation was so great that he became the prestigious Lucasian professor of mathematics at Cambridge University. Babbage, however, did not want to give up his work, so he never actually taught at the university. He simply kept the position until 1839.

■ GEORGE BOOLE (1815–1864)

English mathematician George Boole is sometimes called the father of symbolic (or mathematical) logic. Symbolic logic is a field of mathematics in which logical statements are expressed with numbers and letters.

Boole was born on November 2, 1815, in Lincoln, England. He was the son of a shoemaker who was very interested in mathematics and astronomy. The elder Boole taught his son how to make optical instruments and instructed him in the basics of mathematics. Beyond his father's instruction and a few years at a local school, young Boole received little formal education. His thirst for knowledge, however, continued and he taught himself many subjects considered part of a traditional education. By the age of sixteen, he had studied enough to earn a position as teacher at a Lincoln school. Four years later, he opened his own school.

Boole continued to study mathematics by reading books and periodicals and by 1839, when he was twenty-four years old, he began to submit papers to the *Cambridge Mathematical Journal*. Even though Boole was an amateur mathematician with no formal schooling, his ideas were so original that they received quite a bit of notice. In 1847, Boole wrote his first formal paper. In it, he suggested that logical ideas could be expressed mathematically.

Boole expanded his theory of symbolic logic in his most famous book, *An Investigation into the Laws of Thought, on Which are Founded the Mathematical Theories of Logic and Probabilities*, written in 1854. According to Boole, symbols from algebra could be used to express logical arguments. For example, consider the statements "The door is open" and "The door is

George Boole. (Courtesy of the Library of Congress.)

Mathematics

BIOGRAPHIES

closed." Those two statements could be expressed, Boole said, with the numbers "1" and "0." Then consider the statements "The window is open" and "The window is closed." Those two statements could also be expressed with the numbers "1" and "0." By adding the numerical value of both statements (1 + 1), the relationship would be: "Both the door and the window are open."

In 1849, based on the strength of his work, Boole was asked to take a position as professor of mathematics at Queen's College in Cork, Ireland. He remained at Queen's College until his death in 1864. His death was caused by pneumonia, which developed after Boole gave a lecture while still soaking wet from a walk in the rain.

Boole did receive acclaim during his lifetime. In fact, he was given an award by the Royal Society before he was thirty years old for his contributions to the study of algebra. He also became a Fellow of the Royal Society and was highly regarded as an excellent teacher.

It was after his death, however, that the true impact of his work became apparent. Mathematicians eventually developed a whole new field of mathematics based on Boole's ideas, a field now known as Boolean algebra. An especially important application of Boolean algebra is in the field of computer science, which is entirely based on the binary (two-value) system of George Boole.

■ AUGUSTIN-LOUIS CAUCHY (1789–1857)

French mathematician Augustin-Louis Cauchy is remembered as one of the greatest mathematicians of the nineteenth century. He published more than seven hundred scientific papers during his lifetime that helped advance nearly every field of mathematics. He was also one of the most influential scholars in the French scientific community.

Cauchy lived in France during a period when many governments came to power and fell from power. This political upheaval would greatly influence Cauchy throughout his life. He was born in Paris, the son of a brilliant scholar who was a government official. To escape the terror of the French Revolution (1789), the family moved to the country village of Arcueil. The great French scientists **Pierre-Simon Laplace** (1749–1827; see biography in this chapter) and Pierre Lagrange (1736–1813) also lived in the region and were frequent visitors in the Cauchy home.

Laplace recognized young Cauchy's talent for mathematics and encouraged him to study at the École Polytechnique. After graduation, Cauchy studied engineering at the École de Pontes et Chaussées. He left school in 1809 to work as a military engineer at the Ourcq Canal works and later at the Cherbourg harbor naval base.

In 1815, he returned to Paris and one year later, he was made full professor at the École Polytechnique. At the same time, he became the youngest person to be elected to the Academy of Sciences. Cauchy was twenty-seven years old.

During the 1820s, Cauchy was incredibly productive. In three famous essays, he established one of the first complete theories of complex numbers. He also contributed to the fields of astronomy, mechanics, and physics. To this day, more theorems are named after Cauchy than any other mathematician. In the field of mathematical elastic theory alone, which Cauchy invented, there are sixteen principles named after him.

One of Cauchy's greatest achievements was in clarifying the principles of the calculus. By doing so, he helped establish calculus as a useful branch of mathematics. Cauchy's theories are still taught today to help students understand some of the most basic concepts of calculus.

In 1830, politics disrupted Cauchy's life when Louis-Philippe (1773–1850; king of France 1830–1848) replaced Charles X (1757–1836; king of France 1824–1830) as ruler of France. Cauchy refused to pledge allegiance to the new king and, as a result, he lost his position at the university.

Cauchy fled the country and lived for several years in Italy. In 1833, he moved to Prague, where the royal family was also in exile. While there, he tutored the son of Charles X. In 1838, Cauchy returned to Paris and eventually resumed his work at the university. Because of his important contribution to French science, Cauchy was never again asked to pledge loyalty to the French government.

While Cauchy was a respected mathematician and scientist, he was also known to be a stubborn man who treated his fellow mathematicians very harshly. In some instances, this resulted in a scientist's work being ignored or dismissed by the French scientific community. An example occurred in 1829, when **Evariste Galois** (1811–1832; see biography in this chapter) attempted to present a paper introducing group theory to the French Academy of Sciences. Cauchy virtually ignored the paper and, according to some sources, actually lost the manuscript.

■ JEAN-BAPTISTE JOSEPH FOURIER (1768–1830)

Jean-Baptiste Fourier was a French mathematician who developed a method for analyzing complex periodic motion. An example of such a motion is a wave. His method is known today as Fourier series or Fourier analysis.

Mathematics

BIOGRAPHIES

Fourier was born in Auxerre, France, on March 21, 1768, the ninth child of a tailor and his wife. Both his parents died by the time Fourier was ten years old. He was then sent to a military school operated by Benedictine monks. There he became interested in mathematics and showed quite a proficiency for the subject. He eventually became a teacher at the school.

For a while, Fourier considered becoming a priest because he wanted the opportunity to pursue his mathematics education at the Benedictine seminary in Paris. This was during the aftermath of the French Revolution (1789), however, and he became involved in politics instead. He joined the local Revolutionary Committee, hoping to establish a government free from rule by a monarch (king). During the period after the revolution, one group after another came to power in France. Fourier was sometimes on the winning side, and sometimes on the losing side. As a result, he ended up in jail twice before he was thirty years old.

In 1798, Fourier joined Napoleon I (1769–1821; emperor of France 1804–1815) on an expedition to Egypt. He served as a consultant on construction projects in Egypt and also acted as a diplomat for Napoleon. Fourier soon became an expert on Egyptian culture. When he returned to France in 1801, he began work on a massive description of scientific and cultural discoveries made in Egypt, *Description d' Egypt* (Description of Egypt).

While Napoleon was in power, Fourier held various positions of authority and honor. When Napoleon fell from power, Fourier returned to Paris where he was appointed head of the Statistical Bureau of the Seine.

Jean-Baptiste Fourier. (Courtesy of the Library of Congress.)

Fourier's most important mathematical work began while he was administrator of the French region of Isere. He was interested in trying to explain the conduction of heat using mathematical equations only. In order to do so, he had to develop new mathematical techniques. One of those techniques was the Fourier series.

A Fourier series is used to analyze any kind of periodic behavior, such as the rise and fall of water waves or the flow of an electromagnetic wave through space. Fourier found that such wave patterns could be broken down into a series of simpler patterns. Each of these simpler patterns could then be represented by an easy-to-solve trigonometric function. This technique is widely used today to analyze all types of wave phenomena, including sound waves and light waves.

Fourier also made contributions in many other areas of mathematics, such as probability theory, theory of equations, and error theory. None, however, matched his development of Fourier series. Fourier died in Paris on May 16, 1830, as the result of a disease he had contracted in Egypt many years before.

■ EVARISTE GALOIS (1811–1832)

Evariste Galois. (Reproduced by permission of Corbis-Bettmann.)

Evariste Galois was a French mathematician who led a short, eventful, and tragic life. He was born on October 25, 1811, in Bourg-la-Reine, near Paris. His father was active in politics and served as mayor of his home town. His mother came from a well-known French family, all of whom strongly believed in education for both men and women. She taught her son at home until he reached the age of twelve. Galois was then sent to the College Royal de Louis-de-Grand.

The quality of Galois's education at the College Royal was not very good. One of his teachers, however, did notice that Galois was particularly gifted in mathematics and advised him to study algebra. He quickly mastered most of the basics of mathematics and soon began to tackle some of the most puzzling problems in the field.

By the age of seventeen, Galois had written three papers on his most important theory. He had devised a method to determine whether or not a given equation is solvable. The techniques he developed later became a fundamental part of the field of mathematics now known as group theory.

Galois attempted to have his work published by the Academy of Sciences, however, all three of his papers were either lost or ignored by reviewers. He suffered further professional setbacks when he was refused admittance to the École Polytechnique, the most famous school of mathematics in France. On both occasions, he became involved in arguments with his examiners and was rejected as a student. Eventually he was admitted to the less prestigious École Normale Superieure.

Galois's personal life was also in a shambles. In 1829, his father committed suicide following bitter disagreements with other politicians in his home town. At the time, the entire French nation had fallen into political turmoil. A citizens' revolution had forced King Charles X (1757–1836; king of France 1824–1830) into exile, and placed Louis-Phillippe (1773–1850; king of France 1830–1848) on the throne. Galois was arrested twice for his role in the political upheaval and was sent to jail for six months on the second occasion.

Galois's life ended in Paris on May 31, 1832, during a duel. He was twenty-one years old. The circumstances leading to the duel are still not known. Some people believe it may have involved an unhappy love affair, while others think it was related to politics.

The world eventually found out about Galois's work primarily through the efforts of French mathematician Joseph Liouville (1809–1882). Liouville collected and edited Galois's papers before having them published in 1846 in the *Journal de Mathematiques Pures et Appliquees*. In 1870, a detailed treatment of Galois's theory was published by French mathematician Camille Jordan (1838–1922). These publications brought Galois's work to the attention of the mathematics community and earned him a place in the history of mathematics.

Galois's work was considered brilliant and imagination and his contributions were key in advancing group theory. Group theory eventually pervaded all of mathematics. It also influenced scientific fields, including particle physics. It is impossible to guess the contributions Galois might have made if he had lived into old age.

■ CARL FRIEDRICH GAUSS (1777–1855)

Carl Friedrich Gauss is considered to be one of the greatest mathematicians of all time. He made important contributions not only in mathematics, but also in physics, astronomy, magnetism, and many other scientific fields.

Gauss was born into a poor family in Braunschweig (now Brunswick), Germany, on April 30, 1777. He showed an early genius for mathematics and supposedly was able to make calculations before he could talk. He taught himself to read and showed great aptitude for languages. He also kept random lists of numerical data. Gauss was not only skilled with numbers, he was consumed by them.

Gauss's talent was recognized by his teachers and his parents, who brought him to the attention of the Duke of Brunswick. The duke was so impressed by the fourteen-year-old that he agreed to provide financial support for Gauss's schooling.

Mathematics

BIOGRAPHIES

In 1792, Gauss enrolled at the Collegium Carolinum in Brunswick. He received his doctoral degree from the University of Helmstedt in 1799. For his doctoral dissertation, Gauss completed the proof of the Fundamental Theorem of Algebra. That theorem says that any polynomial equation has at least one solution. The theorem had been widely studied before, but Gauss provided its fully correct interpretation.

Before Gauss received his doctoral degree, in fact, while he was still a teenager, he made many important discoveries. One of his early discoveries is known as the principle of least squares. According to this principle, it is possible to draw a curve that fits a set of data points. Today that principle is widely used to draw curves that describe observations made in an experiment.

In 1801, at the age of twenty-four, Gauss published what some historians regard as his greatest work, *Disquisitiones Arithmeticae* (Arithmetical Researches). This book covers a number of topics in number theory, including research on complex numbers.

After 1800, Gauss turned some of his attention to astronomy. In 1801, the asteroid Ceres was discovered. Astronomers were unable to predict the asteroid's orbit, so Gauss developed a mathematical technique to plot its movements. Astronomers were then able to make repeated observations of the asteroid. Gauss's methods are still used today.

In 1807, Gauss became a professor at the University of Göttingen and director of the university's observatory. He remained at the university until he died in 1855.

Carl Friedrich Gauss. (Courtesy of the Library of Congress.)

About 1820, Gauss started working in the field of geodesy, the measurement of the Earth's surface. He invented the heliotrope, a more accurate surveying instrument than was previously available.

Gauss's work in geodesy led to another mathematical breakthrough. He noticed that lines drawn on the surface of a sphere (like the Earth's surface) had different properties from those drawn on a flat surface. His work in this area influenced his student Bernhard Riemann (1826–1866) to later develop some of the first principles of the theory of relativity.

You, Too, Can Be a Mathematical Genius

The following story, which may or not be true, has often been told about German mathematician Carl Gauss. The math involved, however, is real.

When Gauss was nine years old, his teacher gave his class the problem of adding the integers from 1 to 100. One way to attack the problem is obvious: simply add 1 + 2 + 3 + 4 + 5, and so on, until you reach 100. That method will take some time.

Gauss amazed his teacher by responding with the correct answer in a matter of seconds. He had quickly solved the problem because he noticed a pattern: when the greatest number in the series (100) is added to the least number (1), the result is 101; when the second highest number (99) is added to the second lowest number (2), the result is 101; when the third highest number (98) is added to the third lowest number (3), the result is 101; and so on. Gauss knew that there are 50 pairs of integers between 1 and 100. Therefore, the sum of these integers is 101 × 50 or 5,050.

The pattern Gauss noticed can be used to find the sum of the integers from 1 to any even number, from 1 to 200, for example. Here is how the method works:

Step 1: Add 1 to the even number. 1 + 200 = 201

Step 2: Divide the even number by 2. 200 ÷ 2 = 100

Step 3: Multiply the answers from Step 1 and Step 2. 201 × 100 = 20,100

See if you can use these steps to find the sum of the integers from 1 to 1,000,000.

Gauss was also interested in the application of mathematics to electricity, magnetism, and other fields of physics. He developed a theory explaining the way electric charge is distributed through the surface of an object. That theory is now known as Gauss' Law. In honor of his work in magnetism, the standard unit of measurement of magnetic influence was later named the *gauss*.

■ SOPHIE GERMAIN (1776–1831)

Mathematics

BIOGRAPHIES

Sophie Germain is considered to be France's greatest woman mathematician. Her work in applied mathematics was so original that some call her one of the founders of mathematical physics.

Marie-Sophie Germain was born April 1, 1776, in Paris, France, the daughter of a wealthy merchant who was very involved in politics. After the fall of the Bastille in 1789, the Germain family went into hiding to escape the turmoil of the French Revolution. While in hiding, young Sophie took advantage of her father's immense library, and began to read everything she could. She was especially inspired by the legend of Greek scientist Archimedes (c. 287–212 B.C.), who gave his life for the study of mathematics.

Germain taught herself Greek and Latin, and intensively studied geometry, algebra, and calculus. Her hope was to have a career in mathematics, even though her parents tried to discourage her. At the time, many people believed that the study of mathematics and science was too strenuous for women and would lead to ill-health. Germain was so determined to study that she often sneaked out at night and read by the light of smuggled candles.

Because she was a women, Germaine could not pursue an advanced degree at a university. She did, however, study the lecture notes of friends who were attending the prestigious École Polytechnique, which was the most important school for mathematicians of the day. At the end of a term, students were asked to comment on their professors' lectures. Germain was so intrigued by the lectures of French mathematician Joseph-Louis LaGrange (1736–1813) that she submitted a paper signed "M. le Blanc."

Sophie Germain. (Reproduced by permission of The Granger Collection.)

LaGrange was so impressed by the paper that he contacted the author. He was surprised that "le Blanc" was actually a woman, but the fact did not matter. LaGrange and Germain became friends and the great mathematician provided encouragement to Germain throughout her life.

Germain was also in contact with other famous mathematicians, including **Carl Friedrich Gauss** (1777–1855; see biography in this chap-

Mathematics

BIOGRAPHIES

ter) and Adrien-Marie Legendre (1752–1833). It was during her correspondence with Legendre that she became interested in number theory, a branch of mathematics that focuses on the properties of numbers. As a result, she came up with a partial answer to a famous mathematical problem called Fermat's last theorem.

Because of her work, Germain was accepted into the French scientific community. She was the first woman allowed to attend meetings of the French Academy of Science and the first woman invited to sessions at the Institute of France. In 1816, she was awarded a prize by the French Academy for her theory that explained vibrations in mathematical terms. Winning the prize cemented Germain's place among the great mathematicians of the day.

Germain died in 1831 of breast cancer. For her contributions to the new field of mathematical physics, she was given an honorary doctorate in 1837. Germain was the first woman in history to receive such an honor. Today, she is acknowledged for her talents and her pioneering spirit. In Paris, there is a school named for her, the École Sophie Germain, as well as a street, la rue Germain.

■ FELIX KLEIN (1849–1925)

Felix Klein was a German mathematician whose greatest contribution was in showing how various fields of mathematics were connected to each other. He was also very interested in the history of mathematics and was instrumental in advancing mathematics education.

He was born Christian Felix Klein on April 25, 1849 in Düsseldorf, Germany. Klein attended local schools until 1865, when he entered the University of Bonn. Three years later, at the age of nineteen, he was awarded his doctoral degree. He then did postgraduate study at the universities of Berlin, Göttingen, and Paris. It is believed that Klein first became interested in the possibility of unifying different fields of mathematics while studying under Norwegian mathematician Sophus Lie (1842–1899) in Paris.

In 1872, Klein was appointed professor of mathematics at the University of Erlangen. It was there that he developed the concept of *projective geometry*. Projective geometry is a grand unifying theory that attempts to show how Euclidean and non-Euclidean geometries are not really different from each other but, instead, are different forms of a larger field of mathematics.

Imagine, Klein said, that a three-dimensional figure such as a cube was made of thin wire. Then imagine that a light was shined on that cube. The shadow projected by the cube would be a two-dimensional representation of the three-dimensional figure. Or, from a mathematical stand-

Mathematics

BIOGRAPHIES

point, the Euclidean geometry of the shadow would be a representation of the non-Euclidean geometry of the cube.

Depending on the angle at which the light shines on the cube, shadows of many different shapes could be formed. All of these shadows could be represented by different mathematical systems. Using this approach, Klein showed how many apparently different forms of geometry were really all subsets of a large form—projective geometry.

Klein was very active in the development of mathematics as a profession. He wrote many popular books on the theories and history of mathematics. He also wrote texts on the teaching of mathematics, the most famous being *Elementar Mathematik von Hoheren Standpunkte aus,* or "Elementary Mathematics from an Advanced Standpoint," published in 1908.

After leaving the University of Erlangen, Klein taught at the Technical Institute in Munich from 1875 to 1880. From there, he moved to the universities of Leipzig (1880–1886) and Göttingen (1886–1913). During these years, Klein was involved in the reform of mathematics education. He proposed that the calculus and other advanced fields of mathematics should be introduced into secondary school programs. In 1908, he was elected chairman of the International Commission on Mathematical Instruction.

Klein's physical health was never very strong. He had a tendency to overwork, which affected both his mental and physical well-being. He retired from teaching in 1913 because of poor health, but continued to tutor students at his home. He died at Göttingen on June 22, 1925.

■ PIERRE-SIMON LAPLACE (1749–1827)

Pierre-Simon Laplace, sometimes called the Isaac Newton of France, is known for his many contributions to the fields of mathematical astronomy and probability. His published works and scientific philosophy greatly influenced the mathematics community of the nineteenth century.

He was born into a middle-class family in Normandy, France, on March 23, 1749. Through the financial assistance of neighbors, Laplace was able to attend a Benedictine school between the ages of seven and sixteen. He then entered Caen University to study theology, but his apparent gift for mathematics prompted him to pursue a field in the sciences.

When he was nineteen years old, Laplace was made professor of mathematics at the Military Academy of Paris. In 1773, he became an associate member of the Paris Academy of Sciences. Shortly after, he presented a paper in which he proved that the planetary motions of the solar system are stable. The paper expanded on the work of English physicist and

Mathematics

BIOGRAPHIES

mathematician Isaac Newton (1642–1727). It also earned Laplace the nickname, the "French Newton."

In 1796, Laplace published his *nebular hypothesis,* which stated that the solar system evolved from a rotating cloud of gas. According to Laplace, rings of gas were thrown off the cloud as it cooled. As the rings continued to cool, they condensed to form the planets. This hypothesis became very popular during the nineteenth century, but eventually fell out of favor.

Between 1799 and 1825, Laplace published *Mécanique Céleste (Celestial Mechanics)*. This five-volume series was considered the most important work in mathematical astronomy at the time. In it, Laplace collected all of his previous work on solar stability and expanded on the effect gravitation has on planetary orbits.

While working on problems in mathematical astronomy, Laplace became involved in the science of probability. In 1812, he published Théorie Analytique des Probabilités (Analytical Theory of Probability). Laplace's work included many mathematical innovations. It also contained a philosophy of probability that would dominate the science of probability through the nineteenth century—that probability was a rational method to reliably predict the outcome of an event.

In 1814, Laplace expanded on his philosophy of probability in a popular edition of his 1812 *Théorie Analytique des Probabilité*. Laplace wrote that probability could be used for *the important questions of life* that were nearly impossible to answer. For instance, it was prob-

Pierre-Simon Laplace. (Courtesy of the Library of Congress.)

lematic to determine the reliability of witnesses in a legal trial. According to Laplace, in order to predict the odds that witnesses are telling the truth, one had to consider all the variables. Such variables include: the reputation of the witnesses, the number of witnesses, and whether or not the witness testimonies are contradictory.

In addition to his work in mathematics, Laplace was involved in politics during a time of political upheaval in France. He became a member of the Senate under Napoleon I (1769–1821; emperor of France 1804–1815). Laplace died on March 5, 1827. His last words were: "That which we know is mere trifle, that which we are ignorant of is immense."

"Thus It Plainly Appears"

Pierre-Simon Laplace is believed to be one of the greatest mathematicians of the nineteenth century. His five-volume series, *Mécanique Céleste* (Celestial Mechanics), is considered one of the most important works of science ever produced. It is very interesting to note that Laplace had somewhat of an inflated ego. As a result, he was very reluctant to give credit to scientists who were conducting research at the same time. For instance, French mathematician Joseph-Louis Lagrange (1736–1813) had been studying the stability of the solar system at the same time as Laplace. Laplace, however, did not acknowledge Lagrange in *Celestial Mechanics*. Instead, he suggested that all the results described in the work were his own.

In addition, Laplace's text was very difficult to understand, even for professional scientists and mathematicians. He presented results, but often did not explain how they were obtained. For example, he was known for beginning a sentence with, "It is easy to see...."

The problem was that the results were often not "easy to see." One person who pointed out this fact was the American mathematician Nathaniel Bowditch (1773–1838). Bowditch translated Laplace's work into English and included his own comments. Bowditch wrote: "Whenever I meet in LaPlace with the words 'Thus it plainly appears,' I am sure that hours, and perhaps days, of hard study will alone enable me to discover how it plainly appears."

■ NIKOLAI IVANOVICH LOBACHEVSKY (1792–1856)

Nikolai Lobachevsky was a Russian mathematician considered to be one of the founders of non-Euclidean geometry. Euclidean geometry was established by the Greek mathematician Euclid (c. 330–c. 260 B.C.), and for more than 2,000 years, it was considered the one and only form of geometry. The non-Euclidean geometry developed by Lobachevsky, the Hungarian mathematician Janos Bolyai (1802–1860), and others later became an essential building block for Albert Einstein's (1879–1955) theory of general relativity.

Lobachevsky was born on December 1, 1792, in Nizhny Novgorod (now Gorki), Russia, into a very poor family. His father was a low-ranking

Mathematics

BIOGRAPHIES

government official who died when young Nikolai was seven years old. The family then moved to Kazan, at the edge of Siberia.

When he was fourteen, Lobachevsky received a public scholarship to the university in Kazan. He intended to study medicine, but was greatly influenced by one of his teachers, Johann Bartels. Bartels was a skilled mathematics teacher who had taught the famous mathematician **Carl Friedrich Gauss** (1777–1855; see biography in this chapter). Lobachevsky soon gave up his plans to study medicine and switched to mathematics. He remained at the university in Kazan for the rest of his life; he became a professor there in 1816.

In 1826, Lobachevsky announced that he would challenge Euclid's fifth postulate (rule). According to the postulate, only one parallel line can be drawn through a fixed point outside another line. Unbeknownst to Lobachevsky, Bolyai and Gauss had separately been working on the same postulate. (Gauss had worked on it some twenty years before.) Both were successful in developing an alternative to Euclid's postulate, but neither ever published their results. Bolyai and Lobachevsky knew nothing of the other's work, and although Gauss knew them both, he never introduced them.

Lobachevsky published his findings in 1829 in the *Kazan Messenger*. He expanded on his theory in *Geometrical Researches on the Theory of Parallels* (1840) and *Pangeometric* (1855). Most mathematicians now agree, however, that credit for the discovery of non-Euclidean geometry should probably be divided between Bolyai and Lobachevsky.

Nikolai Lobachevsky. (Courtesy of the Library of Congress.)

In addition to being a gifted mathematician, Lobachevsky was also a skilled administrator. He served as rector of his university, dean of mathematics and physics, librarian, and public official. His university success was amazing, considering he served under the reign of Tsar Alexander I (1777–1825; tsar of Russia 1801–1825). The tsar distrusted science and philosophy and withdrew his support for these subjects. During this period, the university lost some of its best teachers.

Conditions changed when Tsar Nicholas I (1796–1855; tsar of Russia 1825–1855) succeeded Alexander in 1826. Lobachevsky worked to restore the university to its previous high standards. He established rules for sanitary practices to protect the university from the cholera epidemic of 1830. He oversaw the reconstruction of several buildings destroyed by a terrible fire in 1842. He also fought to improve primary and secondary education in the region around Kazan.

Ironically, government officials did not appreciate Lobachevsky's efforts. In 1846, officials removed him from his post at the university without explanation. A public outcry arose, but Lobachevsky never returned to the university. He died at Kazan on February 24, 1856.

■ JULES-HENRI POINCARÉ (1854–1912)

Jules-Henri Poincaré is heralded as one the greatest mathematicians of the nineteenth and twentieth centuries. In particular, he is noted for being one of the last great generalists in mathematics. This means that he was interested in almost every field of mathematics. He is also called one of the great philosophers of science.

Poincaré was born in Nancy, France, on April 29, 1854. His father was a well-known doctor, and his mother was well-educated and very involved with her son's education. His first cousin, Raymond Poincaré (1860–1934), served as president of the French Republic during World War I (1914–18).

As a child, Poincaré's eyesight and muscle coordination were poor. He was, however, a brilliant student with a photographic memory. In 1872, he won a competition among high school students throughout France and entered the famous École Polytechnique in Paris. There he won a number of prizes in mathematics and earned a reputation as a "monster of mathematics."

After graduation, he enrolled at the École Nationale Supérieure des Mines, where he received his doctoral degree in 1879. In 1881, he accepted an appointment at the Université de Paris, where he taught for the rest of his life.

Poincaré's lifetime productivity was amazing. He wrote on topics such as number theory, probability theory (see essay "The Rise of Probability Theory" in this chapter), and algebraic geometry. One of his most

Jules-Henri Poincaré (Reproduced by permission of Hulton-Deutsch Collection/Corbis.)

Mathematics

BIOGRAPHIES

Mathematics

BRIEF BIOGRAPHIES

famous contributions to mathematics was his discovery of functions known as *automorphic functions*. An automorphic function is one that does not change when put through certain types of transformations. He showed how automorphic functions could be used to solve a number of different kinds of mathematical problems. In 1887, Poincaré was elected to the French Academy of Sciences in recognition of his mathematical achievements.

Another highpoint of 1887 was when King Oscar II of Sweden (1829–1907; king of Sweden 1872–1907) offered a prize for a solution to the n-body problem. The solution would mathematically explain how three or more massive bodies in space are influenced by each other's gravity. The problem is enormously difficult and has never been completely solved. Poincaré, however, developed a partial solution that won him the Swedish prize. He was also made a knight of the French Legion of Honor.

Poincaré's work on orbital motion led him to create a new field of mathematics, known as topological dynamics. He was also among the scientists at the end of the nineteenth century who were expanding the study of time and space. Before Albert Einstein (1879–1955) released his theory of relativity, Poincaré was independently developing similar theories.

Later in his career, Poincaré turned his attention to writing books that would explain science and mathematics to the general public. One of his philosophies was that creativity was an essential component of scientific progress. Such philosophies brought him acclaim inside and outside the scientific community.

In 1908, Poincaré was elected to the Académie Française. This is the highest honor in the French literary world. By the end of his life, Poincaré had written more than thirty books and five hundred scholarly papers. He died suddenly on July 17, 1912, in Paris.

▲ BRIEF BIOGRAPHIES

▲ EUGENIO BELTRAMI (1835–1899)

Beltram was an Italian mathematician who showed the connection between Euclidean geometry and the non-Euclidean geometry developed by **Nikolai Lobachevski** (see biography in this chapter). Beltrami also applied the principles of mathematics to physical problems, such as those of fluid motion and elasticity.

Mathematics

BRIEF BIOGRAPHIES

▲ ENRICO BETTI (1823–1892)

Betti was an Italian mathematician who made important contributions in the fields of algebra and topology (the study of surfaces). He developed a method for solving quintic equations (equations with a variable raised to the fifth power). Betti was also active in politics and held many important political positions during his life.

▲ BERNHARD PLACIDUS JOHANN NEPOMUK BOLZANO (1781–1848)

Bolzano was a Czech philosopher, mathematician, and theologian whose special interest was the subject of mathematical infinities. He was an ordained Roman Catholic priest who wrote about philosophy and politics, as well as mathematics. Some of his writings were banned by the government. Most of his work in mathematics was not published until after his death, so he was not recognized in this field during his lifetime.

▲ MARY EVEREST BOOLE (1832–1916)

Boole was a mathematics educator who invented the term string geometry to describe the principles of angles and shapes. She worked as a librarian at Queens College, the first women's college in England, since women were not then permitted to teach there. Boole became interested in psychology of the learning process. In 1904 she wrote a book, *Preparing the Child for Science,* that had a major impact on progressive education in England and the United States.

▲ ADA AUGUSTA BYRON, COUNTESS OF LOVELACE (1815–1852)

Byron was the daughter of poet Lord Byron and a successful mathematician and pioneer of computer science. She taught herself basic mathematics and was later helped with advanced topics by Augustus De Morgan, the first professor of mathematics at the University of London. Byron is best known for her work with mathematician **Charles Babbage** (see biography in this chapter). She translated Babbage's work and included her own notes and comments about his manuscripts. Mathematicians paid relatively little attention to her work during her lifetime. But she is now widely admired, and an early computer programming system, ADA, was named in honor.

▲ MORITZ BENEDIKT CANTOR (1829–1920)

Cantor was a German historian of mathematics. Between 1880 and 1908, he published a four-volume history of mathematics, *Vorlesungen Über Geschichte der Mathematik* (*Lectures on the History of Mathematics*). That

Mathematics

BRIEF BIOGRAPHIES

work is still considered one of the finest histories of mathematics available. It covers the period from the earliest humans through the eighteenth century. Cantor was also editor of the prestigious journal *Zeitschrift für Mathematik und Physik*.

▲ MICHEL CHASLES (1793–1880)

Chasles was a French mathematician known for his contributions in the fields of algebraic and projective geometry. He wrote an important book on the methods used in geometry, still in use today. Chasles also developed the theory of cross ratios at about the same time as did August Möbius.

▲ WILLIAM KINGDON CLIFFORD (1845–1879)

Clifford was an English mathematician best known for his work in non-Euclidean geometry and topology (the study of surfaces). He made important contributions to the mathematical analysis of matter, energy, and space. Some of his conclusions were used by Albert Einstein in his general theory of relativity. Clifford died at the age of thirty-five, apparently from stress induced by severe overwork.

▲ ANTOINE-AUGUSTIN COURNOT (1801–1877)

Cournot was a French mathematician who first applied the methods of mathematics to the study of economics. His earliest work was in the field of mechanics, a branch of physics. He later became professor of mathematics at the universities of Lyon and Grenoble. There he became interested in mathematical economics. Many of the methods developed by Cournot, including the use of the term market, are still in use today.

▲ AUGUST LEOPOLD CRELLE (1780–1855)

Crelle was a German mathematician who, in 1826, founded the first journal devoted entirely to mathematics. It is known today as *Crelle's Journal*. He also published a number of mathematical textbooks and many editions of multiplication tables. His journal made possible the publication and circulation of many important early papers in mathematics.

▲ AUGUSTUS DE MORGAN (1806–1871)

De Morgan was an English mathematician whose special interest was in the field of logic. His work helped encourage the development of mathematical logic as a separate discipline. De Morgan also played an important role in making mathematics a profession in England. He worked to make mathematics education more generally available to university students, such as those planning on careers in science and technology.

Mathematics

BRIEF BIOGRAPHIES

▲ PHILBERT MAURICE D'OCAGNE (1862–1938)

D'Ocagne was a French mathematician who created the field of nomography. Nomography is a method that uses graphs to solve equations. The method is especially useful for solving problems in the field of engineering. D'Ocagne's work was translated into more than twelve languges.

▲ HERMANN GÜNTHER GRASSMAN (1809–1877)

Grassman was a German-Polish mathematician best known for his work on the calculus of vectors. He also developed a form of geometrical algebra that used geometrical symbols rather than numbers. At the age of fifty-three, he became discouraged because he had received so little recognition for his work in mathematics. He decided to study the Sanskrit language instead. He eventually wrote a Sanskrit dictionary that is still respected today.

▲ DAVID HILBERT (1862–1943)

Hilbert was a German mathematician widely regarded as one of the most important mathematicians of the nineteenth century. His primary field of interest was geometry, where he made some of the most important advances since the time of Greek mathematician Euclid (c. 330–c. 260 B.C. He also studied the use of mathematics in solving problems in physics. Some of his work in this area was similar to that conducted by Albert Einstein at about the same time. Hilbert also served with distinction in many mathematical organizations.

▲ GEORGE WILLIAM HILL (1838–1914)

Hill was an American mathematician and physicist whose special interest was the application of mathematics to problems in astronomy. He calculated the orbits of various astronomical orbits, such as that of the Moon and the moons of Jupiter. He also studied the interaction of three and four bodies, one of the most difficult of all physical problems. He was elected a Fellow of the Royal Society and was given that association's Copley Medal. These were important accomplishments at a time when most fields of science in the United States was not yet well developed.

▲ SOFYA VASILYEVNA KOVALEVSKAYA (1850–1891)

Kovalevskaya was a Russian mathematician who made important contributions to the study of partial differential equations, celestial mechanics, and the properties of rigid bodies. She worked at a time when opportunities for women in science and mathematics were very limited. But she impressed her colleagues with her talents and made a reputation for her-

Mathematics
BRIEF BIOGRAPHIES

self. She died from influenza and pneumonia at the age of forty-one, just as she was reaching the height of her career.

▲ EMILE MICHEL HYACINTHE LEMOINE (1840–1912)

Lemoine was a French mathematician who made contributions in the field of geometry. His work in mathematics was characterized by short, elegant proofs that were, however, sometimes difficult to follow. He was also an accomplished musician, a civil engineer, and editor of an important mathematical journal.

▲ JOHANN BENEDICT LISTING (1808–1882)

Listing was a German mathematician who first used the term topology for the study of surfaces. He also studied the properties of a surface known as a Möbius strip, discovered by August Möbius in 1865. Listing also studied the application of mathematics to physical problems, especially those in the field of optics (the study of light).

▲ ANDREY ANDREYEVICH MARKOV (1856–1922)

Markov was a Russian mathematician who developed an important mathematical tool used in many fields of mathematics and science today. The Markov process is a method used to predict the future outcome of some process that depends only on the last preceding step. The tool is used in the analysis of a wide range of problems ranging from the diffusion of gases to traffic problems.

▲ AUGUST FERDINAND MÖBIUS (1790–1868)

Möbius was a German mathematician who is best remembered today for an unusual geometric shape he invented. That shape is now called a Möbius strip. A Möbius strip has only one side and two edges. Möbius was also very interested in astronomy, which he taught at the University of Leipzig. He also supervised the construction of the university's observatory.

▲ GASPARD MONGE (1746–1818)

Monge was a French mathematician who invented the discipline known as descriptive geometry. He also was a pioneer in the fields of analytical geometry and projective geometry. Monge was also involved in the creation of the metric system and the founding of the famous École Polytechnique in Paris.

▲ ELIAKIM HASTINGS MOORE (1862–1932)

Moore was an American mathematician whose primary contribution was helping to develop a sound mathematical profession in the United States.

Under his leadership, the University of Chicago's department of mathematics became one of the finest of its kind in the world. He served as president of the American Mathematical Society in 1901 and edited the *Transactions of the American Mathematical Society* from 1899 to 1907. His special areas of interest in mathematics were algebra and group theory.

◼ MORITZ PASCH (1843–1930)

Pasch was a German mathematician who rediscovered many ideas from the work of Greek mathematician Euclid (c. 330–c. 260 B.C.) on geometry. These ideas had laid hidden for more than two thousand years until Pasch's work. Pasch argued that people who studied geometry relied too heavily on their own intuition. He said that more formal proofs of geometric concepts were needed. Pasch's work was to be very influential in David Hilbert's later analysis of Euclidean geometry.

◼ BENJAMIN PEIRCE (1809–1880)

Peirce is generally regarded as the first American research mathematician. He was professor of mathematics and astronomy at Harvard University from 1833 to 1880. He also served as superintendent of the Coast Survey from 1867 to 1874. His special field of interest in mathematics was a study of various possible systems of algebra. His book *Linear Associative Algebra* dealt with that subject. It was published the year after his death.

◼ SIMÉON-DENIS POISSON (1781–1840)

Poisson was a French mathematician, physicist, and astronomer. He was especially interested in the application of mathematical theory to the solution of problems in electricity, magnetism, heat, sound, and other areas of physics. He studied the path of projectiles through the air, based on the work of Gaspard Coriolis. The work for which he is best known in mathematics is his analysis of the theory of probability.

◼ MARY SOMERVILLE (1780–1872)

Somerville was a Scottish mathematician and science writer. She published original papers in physics, as well as an 1831 translation of Pierre Laplace's *Mechanique Celeste (Celestial Mechanics)*. She was one of the few female mathematicians of her time. She became interested in the subject while studying Greek mathematician Euclid (c. 330–c. 260 B.C.) with her younger brother's tutor. She was a strong advocate of women's education and published her own book, *Finite Differences,* as well as several other books on popular science. Oxford University honored her by naming one of its colleges after her.

Mathematics

BRIEF BIOGRAPHIES

Mathematics
BRIEF BIOGRAPHIES

▲ KARL GEORG CHRISTIAN VON STAUDT (1798–1867)

Staudt was a German mathematician and astronomer who did important work in the branch of mathematics known as projective geometry. He discovered a method for solving quadratic equations geometrically. He also determined the orbit of a comet while studying for his doctorate.

▲ JACQUES-CHARLES-FRANÇOIS STURM (1803–1855)

Sturm was a Swiss mathematician and physicist best known for his work with differential equations. Sturm also did research in physics. He made the first accurate measurement of sound in water. He also studied the diffusion of heat, although he never published the results of his investigations. The mathematical analysis that he did, however, later proved very helpful to other researchers.

▲ PETER LUDWIG MEJDELL SYLOW (1832–1918)

Sylow was a Norwegian mathematician whose fame is based almost entirely on a single ten-page paper published in 1872. The paper deals with three theorems in finite group theory. These theorems form the foundation of that field. They are still used in much of the work done in the area of finite group theory. Based on Sylow's reputation, the great mathematician Sophus Lie (1842–1899) arranged to have a special chair created for him at Christiana University. Sylow taught there from 1898 until his death two decades later.

▲ PETER GUTHRIE TAIT (1831–1901)

Tait was a Scottish mathematician and natural philosopher. He work with William Thomson (later Lord Kelvin) on the theory of knots. He tried to show that molecules are simply knots formed in space. Tait also studied thermal conductivity and thermoelectricity. He is perhaps best remembered for having been selected for a chair at the University of Edinburgh by beating out **James Clerk Maxwell** (1831–1879; see biography in Physical Science chapter), later to become one of the greatest mathematical physicists of the eighteenth century.

▲ JOHN VENN (1834–1923)

Venn was an English mathematician and minister best known today for the diagrams he invented to represent sets and their intersections. In addition to set theory, Venn studied logic and probability theory. He also lectured on Moral Science at Cambridge University. Venn also co-authored a history of Cambridge University and invented a machine for bowling cricket balls.

▲ VITO VOLTERRA (1860–1940)

Volterra was an Italian mathematician whose most important work was in the area of integral (whole-number) equations. He demonstrated methods by which mathematics could be applied to problems in physics and the predator-prey relationship in biology. In 1931, he refused to take an oath of allegiance to the fascist Italian government. As a result, he was forced to abandon his position at the University of Rome and leave Italy for the rest of his life.

▲ LÉON WALRAS (1834–1910)

Walras was a Dutch mathematician who helped found the mathematical study of economics. His book, *Elements of Pure Economics,* showed how economic ideas could be expressed in mathematical terms. The book also advocated state nationalization of land and abolition of taxes. His ideas became popular in the United States and Italy, but were rejected in Great Britain and France, at least until after his death.

▲ GRACE EMILY CHISHOLM YOUNG (1868–1944)

Young was an English mathematician who studied spherical trigonometry and calculus. Young's family talked her out of studying medicine, so she took up mathematics instead. She and her husband, mathematician William Young, published more than two hundred papers and several books. Some of these works were published under his name only since women were still not accepted in the fields of mathematics and science.

▲ HIERONYMOUS GEORG ZEUTHEN (1839–1920)

Zeuthen was a Danish mathematician best known for his studies of medieval and ancient Greek mathematics. He also did some original research in mathematics, largely in the field of algebraic geometry. Some of his most important contributions were not fully appreciated until many years after his death.

RESEARCH AND ACTIVITY IDEAS

- Where does your community stand in the debate over mathematics education? Are the math classes in your local schools aimed primarily at consumer education or teaching college preparatory math? Interview teachers and other school officials who can answer this question for you. Then, write an article for the school newspaper reporting your results.

Mathematics

FOR MORE
INFORMATION

- One of the most interesting branches of probability theory is known as game theory. Find out what the basic ideas of game theory are. Write a lesson plan that would explain how you would teach about game theory to one of your friends.

- Mathematicians often work with concepts that, at first glance, seem to have little or no impact on the real world. Some examples are the imaginary number known as i ($\sqrt{-1}$) and figures from topology, such as the Möbius strip and the Klein bottle. Choose one of these examples to study in more detail. Write a report explaining the mathematics behind it. Then, describe any practical applications that the example may have in daily life.

FOR MORE INFORMATION

Books

Barry, P. D. *George Boole: A Miscellany.* Cork: Cork University Press, 1969.

Bell, Eric Temple. *The Development of Mathematics.* New York: McGraw-Hill, 1945.

Bidwell, James K., and Robert G. Clason. *Readings in the History of Mathematics.* Washington, D.C.: National Council of Teachers of Mathematics, 1970.

Boyer, Carl B. *A History of Mathematics.* New York: Wiley, 1968.

Bühler, W. K. *Gauss: A Biographical Study.* New York: Springer-Verlag, 1981.

Fauvel, John, Raymond Flood, and Robin Wilson, eds. *Mobius and His Band: Mathematics and Astronomy in Nineteenth-Century Germany.* Oxford: Oxford University Press, 1993.

Grattan-Guinness, I. *Joseph Fourier, 1768–1830: A Survey of His Life and Work.* Cambridge, MA: MIT Press, 1972.

Herivel, John. *Joseph Fourier: The Man and the Physicist.* Oxford: Clarendon Press, 1975.

Hyman, Anthony. *Charles Babbage: Pioneer of the Computer.* Princeton, NJ: Princeton University Press, 1983.

Kennedy, Hubert C. *Peano: Life and Works of Giuseppe Peano.* Dordrecht, The Netherlands: Kluwer Academic Publishers, 1980.

Kline, Morris. *Mathematical Thought from Ancient to Modern Times.* New York: Oxford University Press, 1972.

National Council of Teachers of Mathematics. *A History of Mathematics Education in the United States and Canada.* Washington, D.C.: National Council of Teachers of Mathematics, 1970.

MacHale, P. D. *George Boole: His Life and Work.* Dublin: Boole Press, 1985.

Moseley, Maboth. *Irascible Genius: A Life of Charles Babbage, Inventor.* London: Hutchinson, 1964.

Petsinis, Tom. *The French Mathematician.* New York: Walker & Company, 1998.

Rigatelli, Laura Toti. *Evariste Galois 1811–1832.* Translated by John Denton. Berlin: Birkhauser Verlag, 1996.

Swade, Doron. *Charles Babbage and His Calculating Engines.* London: Science Museum, 1991.

Web sites

Carl Friedrich Gauss. [Online] http://www.geocities.com/RainForest/Vines/2977/gauss/g_bio.html (accessed on February 26, 2001).

"Charles Babbage." *The MacTutor History of Mathematics Archive.* [Online] http://www-groups.dcs.st-and.ac.uk/~history/Mathematicians/Babbage.html .

"Evariste Galois." *The MacTutor History of Mathematics Archive.* [Online] http://www-history.mcs.st-andrews.ac.uk/history/Mathematicians/Galois.html (accessed on February 26, 2001).

"Felix Christian Klein." *The MacTutor History of Mathematics Archive.* [Online] http://www-history.mcs.st-andrews.ac.uk/history/Mathematicians/Klein.html (accessed on February 26, 2001).

"George Boole." *The MacTutor History of Mathematics Archive.* [Online] http://www-history.mcs.st-andrews.ac.uk/history/Mathematicians/Boole.html (accessed on February 26, 2001).

"Jean Baptiste Joseph Fourier." *The MacTutor History of Mathematics Archive.* [Online] http://www-groups.dcs.st-and.ac.uk/~history/Mathematicians/Fourier.html (accessed on February 26, 2001).

"Jules Henri Poincaré." *The MacTutor History of Mathematics Archive.* [Online] http://www-history.mcs.st-andrews.ac.uk/history/Mathematicians/Poincare.html (accessed on February 26, 2001).

"Nikolai Ivanovich Lobachevsky." *The MacTutor History of Mathematics Archive.* [Online] http://www-history.mcs.st-andrews.ac.uk/history/Mathematicians/Lobachevsky.html (accessed on February 26, 2001).

A Note about George Boole. [Online] http://www.advoool.com/Misc/boole.html. (accessed on February 27, 2001).

"Pierre Laplace." *The MacTutor History of Mathematics Archive.* [Online] http://www-history.mcs.st-andrews.ac.uk/history/Mathematicians/Laplace.html (accessed on February 26, 2001).

chapter four Physical Science

Chronology **212**
Overview **213**
Essays **216**
Biographies **265**
Brief Biographies **285**
Research and Activity Ideas **293**
For More Information **294**

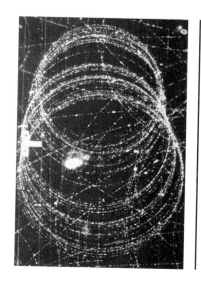

CHRONOLOGY

1800 Italian physicist Alessandro Giuseppe Volta invents the first battery.

1803 English chemist John Dalton proposes the first modern atomic theory.

1812 French naturalist Georges Cuvier proposes the theory of catastrophism, which says that the world has been swept by a number of major catastrophes in its history. Each catastrophe, he says, wipes out all living organisms, which are then replaced by plants and animals of entirely new kinds.

1820 Danish physicist Hans Christian Oersted finds that an electric current can produce a magnetic field. The discovery forms the basis of many modern electromagnetic devices.

1831 English chemist Michael Faraday and American physicist Joseph Henry invent the first electric generators and electric motors.

1837 Swiss-American geologist Jean Agassiz suggests that Earth has been covered a number of times in the past by massive ice sheets, a theory that would not be accepted for some twenty-five years.

1838 British astronomer Thomas Henderson makes the first determination of the distance to a star, Alpha Centauri. At about the same time, German astronomer Friedrich Wilhelm Bessel measures the distance to another star, 61 Cygni.

1846 French astronomer Urbain-Jean-Joseph Leverrier and English astronomer John Couch Adams discover the planet Neptune at almost the same time.

1865 English physicist James Clerk Maxwell develops a set of mathematical equations that shows the relationship between electrical and magnetic fields.

1869 Russian chemist Dmitri Ivanovich Mendeleyev develops the first modern version of the periodic table.

1887 American physicist Albert Michelson and American chemist Edward Morley carry out one of the most famous experiments in scientific history. The experiment fails to detect the presence of an ether, long thought by scientists to permeate the universe.

1888 German physicist Rudolph Hertz first produces and detects radio waves.

1894 British scientists John William Strutt (Lord Rayleigh) and William Ramsay isolate argon, the first in a series of inert gases.

1895 German physicist Wilhelm Conrad Roentgen discovers X rays.

1896 French physicist Antoine-Henri Becquerel finds that certain materials give off X-ray-like radiation, a process later named radioactivity.

OVERVIEW

Chapter Four

PHYSICAL SCIENCE

The basic principles underlying much of physics had largely been developed prior to the nineteenth century through the work of Galileo Galilei (1564–1642), René Descartes (pronounced DAY-cart; 1596–1650), Isaac Newton (1642–1727), and others. Principles such as reliance on objectivity and reason to analyze and explain natural phenomena, modern experimental methods, and Newton's mechanical model of the universe, which held that the behavior of physical objects could be deduced according to certain basic force laws (beginning with his laws of motion; what we now refer to as the laws of physics), were widely accepted among nineteenth century scientists.

But the atomic theory, the cornerstone of modern chemistry, did not appear until 1803. And the basic ideas on which modern earth sciences are based were not to be developed until well into the nineteenth century. Finally, observational astronomy had been an active field of research since the earliest days of human society. But developments in technology during the 1800s were to lead to important breakthroughs in this branch of science also.

Physics

The nineteenth century had barely opened when an exciting discovery in the field of electricity was announced. In 1800, the Italian physicist Alessandro Volta (1745–1827) showed that electricity could be produced by a chemical reaction and made to flow in a current. Before long, physicists were learning much more about the way electrical currents flow and how such currents are associated with magnetism. Hans Christian Oersted (1777–1851), a Danish scientist, noticed that flowing electricity created magnetic effects, which was followed by the discovery by Scottish scientist Michael Faraday (1791–1867) that moving magnets could create electric currents. With these discoveries, the basic principles behind a host of modern devices, such as the electric motor and electric generator, were soon well known. The mathematically precise theory needed to explain the connection between electricity and magnetism, one of the great accomplishments in physics of the century, was announced by the English physicist James Clerk Maxwell (1831–1879) in 1862.

Physical Science

OVERVIEW

James Clerk Maxwell also contributed to our understanding of light by showing that it was a form of electromagnetic energy. As with any good theory, Maxwell's electromagnetic theory suggested a number of experiments. One of the most productive of these experiments was carried out by the German physicist Heinrich Hertz (1857–1894). Hertz showed that a vibrating electrical field was able to produce a wave in that would travel through space, just as Maxwell had predicted. That wave, in turn, was then able to generate a new electric current some distance away. Hertz's discovery laid the foundation for radio, television, and a number of related devices.

Research on heat also produced a number of interesting new discoveries. Scientists investigated the way that heat engines were capable of doing work, either pumping water from flooded mines or turning a huge flywheel, which in turn could power smaller machines. French scientist Sadi Carnot (1796–1832) published his theories about ideal, perfectly efficient engines early in the century. The focus on the conversion of heat to mechanical work led English scientist James Joule (1818–1889) to investigate the reverse process—namely, the conversion of mechanical work to heat. Studies like these led to a clearer understanding of the nature of energy, which can take many forms, one of which is heat. The studies also resulted in the concept we now know as the Law of Conservation of Energy. This law states that, under ordinary circumstances, energy can be neither created nor destroyed; it only changes from one form to another.

Chemistry

The nineteenth century saw the blossoming of chemistry as a science. It developed from a somewhat primitive, poorly organized field of research to a rich and fruitful line of study. In some ways, the most important event was one that occurred shortly after the century started. In 1803, English chemist John Dalton (1766–1844) proposed the atomic theory of chemical elements. In this theory chemical elements were each composed of atoms, and atoms combined in fixed ratios to make chemical compounds (what are now called molecules). According to Dalton, the atoms of different elements differed by weight, so the atoms of any one element all had the same weight and the molecules of any distinct chemical compound also had the same weight, because the proportions of the atoms in the compound were fixed. Dalton's theory has changed somewhat since that time; for example, it is now known that various atoms of an elements can have different weights; nonetheless Dalton's theory has remained the basic concept on which chemistry is based.

The nineteenth century was also one of discovery in the field of chemistry. Fifty-one of the approximately one hundred chemical elements were discovered in the 1800s. These elements ranged from some of the most common and widely used, such as calcium and magnesium, to some of the rarest and most unusual, such as ytterbium and praseodymium.

As the century developed, there was a growing desire to find a way to bring some kind of organization to this collection of elements. In 1869, Russian chemist Mendeleyev (1834–1907) finally solved this problem. He proposed the periodic law of elements. That law said that the properties of the chemical elements are regular functions of their atomic weights. The law allowed Mendeleev to construction an organizational chart of the elements that is still familiar to science students.

Physical Science

OVERVIEW

Earth Sciences

One of the fundamental issues with which earth scientists dealt during the nineteenth century was the age of Earth. In 1800, most scientists believed in the Biblical story of creation. That story could be interpreted to say that Earth was no more than about six thousand years old. But evidence flooded in during the next century that made that timeline impossible.

Instead of relying on the Bible to explain the geologic structures on Earth, scientists had to search elsewhere to develop our planet's story. Two great theories of the Earth's development were proposed. In one, catastrophism, the Earth was described as having gone through one great disaster after another. In the other, uniformitarianism, the Earth's processes were seen as having remained much the same over very long periods of time. The discovery of fossils and evidence of extended ice ages seemed to show that Earth was very old and created new problems for earth scientists to solve.

Astronomy

Scientists who study outer space depend very much on the instruments and techniques they have to work with. During the nineteenth century, there was an explosion of new methods for studying the planets, stars, nebulae, and other parts of outer space. For example, the invention of the spectroscope gave astronomers a new technique for analyzing star light. It allowed them to begin estimating the size, age, and movement of stars that appear no more than a dot of light in the sky. German astronomer Friedrich Bessel not only cataloged the positions of more than three thousand stars, he was among the first astronomers to calculate the distance to a star with a fair amount of accuracy. With Bessel's calculations, astronomers realized that the universe was much bigger than they had originally believed.

Conclusion

For many scholars, the physical sciences had reached their peak of development in the nineteenth century. In fact, the great physicist Albert

Physical Science

ESSAYS

Michelson (1852–1931) made a speech toward the end of the century in which he said that all of the great questions of interest to science had been solved. The twentieth century, he said, held nothing more interesting than calculating numbers to the fifth and sixth decimal place. As exciting as the nineteenth century had been, Michelson was as wrong as a prophet of science had ever been.

ESSAYS

DEVELOPMENT OF THE CONCEPT OF ENERGY

Overview

The concept of energy is fundamental to the understanding all physical motion, whether in nature or resulting from human-made technology. The simplest definition of energy is the capacity to do work; that is, the ability to exert a force over a distance. Energy can reside in many places—in a waterfall, in the wind, in the muscles of the human body, in coal that can be burned, or in the light of the sun. Humans use energy to operate machines, to provide heat and electricity for homes and offices, and to drive many forms of transportation.

While humans have used energy since the earliest times, it was not until the nineteenth century that scientists understood that energy exists in several forms (such as heat, electricity, light, sound, kinetic energy or motion, chemical energy, or nuclear energy) and established most of the physical laws that govern energy. One of the most important nineteenth-century discoveries about energy was that it can be converted from one form to another; and that it can not be created out of nothing nor can it be reduced to nothing. This is the law of the conservation of energy, also known as the first law of thermodynamics.

Background

There are two types of energy that an object can have. Kinetic energy is the energy of motion. Forward motion, turning, spinning, and vibration are all examples of kinetic energy. Potential energy is energy waiting to be released. For example, if a rock is picked up off the ground and held at arm's length, the rock can be dropped, releasing its kinetic energy. While the rock is held suspended above the ground it has potential energy.

Kinetic and potential energy manifest themselves in six basic types of energy: (1) mechanical, (2) thermal, or heat, (3) light, or radiant energy,

Words to Know

energy: The capacity to do work; the ability to exert a force over a distance.

force: In scientific terms, force is what makes objects change their motion.

kinetic energy: The energy of motion.

law of the conservation of energy: Also called the first law of thermodynamics; states that energy can be converted from one form to another; but that it can not be created out of nothing nor can it be reduced to nothing.

potential energy: Energy waiting to be released.

"vis viva": A seventeenth-century scientific term to describe an object's potential for motion, and its ability to exert force on another object.

work: The ability to exert a force over a distance.

(4) chemical, (5) electrical, and (6) nuclear. All were identified and explored in the nineteenth century, with the last of them, nuclear, just being on the threshold of definition when the century ended. As scientists experimented with these various energy forms, they began to understand the connections among them and found evidence that energy never disappears, it is only transformed from one type into the other. Thus, they developed the law of conservation of energy.

Stated another way, the law of conservation of energy says that the amount of energy used to initiate a change is always the same as the amount of energy detected at the end of the change. The energy may be transformed into a number of forms, such as light energy and thermal energy, during the reaction, but if you measure how much of each type of energy results from the reaction it will equal the amount of energy that went into the action. For example, when a moving ball hits another ball, the first ball's kinetic energy is transferred to the second ball, while a small amount of the first ball's energy is also converted into thermal energy. If you were to measure the sum of the thermal and kinetic energy that existed after the collision it would equal the amount of kinetic energy in the first ball.

Physical Science

ESSAYS

The progression of scientific study that led to our modern understanding of energy began in the seventeenth century. At that time, when scientists discussed an object's potential for motion, and its ability to exert force on another object, they referred to these qualities as the object's "vis viva" or "life force." Physicists such as Isaac Newton (1642–1727) mostly considered simple mechanical systems, like billiard balls or planets in orbit when studying the laws of motion. Seventeenth-century Dutch scientist Christiaan Huygens (1629–1695), studying colliding objects, concluded that the force of the objects after the collision changed but was not lost: This realization helped establish the idea that energy can be transformed from one type to another.

English physicist and physician Thomas Young (1773–1829) first coined the term "energy" in 1807. He established the definition of energy as the ability to do work. Interestingly, scientists paid almost no attention to Young's use of the term energy. It was largely ignored until the 1850s, when it was once again introduced and became popular. The long delay in adopting this term suggests that scientists did not really understand the concept well enough when Young first proposed the term.

In the nineteenth century, as investigations into the properties of heat, electric charge, light, and chemical bonds deepened, it became clearer to scientists that energy can take many different forms and it may convert from one form into another. In 1800, for example, the Italian physicist Alessandro Volta (1745–1827) demonstrated that the energy produced from chemical reactions could be transformed into electricity. Twenty years later, the Danish physicist Hans Christian Oersted (1777–1851) showed that an electric current produces a magnetic field. In the 1820s, experiments by the American scientist Benjamin Thompson (also known as Count Rumford; 1753–1814) showed that heat was generated by mechanical motion, and was not a liquid as previously believed. Thompson's calculations to quantify the energy conversion were, however, not accurate.

Beginning around 1839 English scientist James Joule (1818–1888) conducted a series of experiments that proved the equivalence of mechanical work and heat. Joule was the first to precisely quantify how much heat is produced by mechanical work, affirming Thompson's experiments that demonstrated that heat is a form of energy. German physician Julius Robert Mayer (1814–1878) reached the same conclusion at about the same time based on observations of the living body. Both Joule and Mayer provided a mechanical explanation of the nature of heat in which heat was the energy of motion: kinetic energy.

More importantly, experiments in the nineteenth century began to show that there was more than just a transformation of energy. Scientists began to realize that the amount of energy lost in one form was always the same as the amount of energy gained in another form. That is, the total amount of energy remained constant whenever it changed from one form to another.

A number of researchers contributed to the statement of this concept. Both Joule and Mayer contributed greatly to the law of conservation of energy. They certainly deserve credit for taking part in the law's discovery. But it was probably the work of German physicist Hermann von Helmholtz (1821–1894) that finally convinced the scientific world of the law's truth. Helmholtz was one of the most widely admired of all scientists of his day, and he provided the most comprehensive treatment of the concept of conservation of energy in an 1847 paper. With Helmholz's mathematical rendering of the theory, most scientists finally accepted the law of conservation of energy as true.

Physical Science

ESSAYS

Impact

The practical applications of the understanding of energy and energy transformation and conservation led to the development or the improvement of countless devices and machines. The improvements contributed

The Barrage de la Rance hydroelectric dam, located on the Rance River. Improved understanding of the concept of energy led to the development of hydroelectric power plants. (© Jim Sugar Photography. Reproduced by permission of Corbis-Bettmann.)

Physical Science

ESSAYS

to the success of the Industrial Revolution in the nineteenth century. Understanding the laws governing energy allowed for the modification and improvement of steam engines and the instruments measuring the limits of those engines. Consequently, the need to improve the efficiency in the energy cycle with steam led to the greater efficiency of the steam turbine. Later, in the quest of ever-increasing efficiency, adopting a chambered engine to provide energy led to the development of the internal combustion engine. This led to advances in heating, ventilating, and refrigerating systems, and power plant applications of energy designed to provide greater yield led to natural gas, electric, and hydroelectric systems.

Countless everyday devices rely on the transformation of energy from one form to another. Electricity, for example, is transformed into many useful forms. In electric stoves and radiators an electrical current flows through metallic coils and is converted into heat. Lightbulbs turn electricity into light (and heat). In a television set an electrical beam strikes a thin layer of chemicals on the television screen, causing them to glow and produce light, which forms a picture. All of these devices rely on an understanding of the principles of energy transformation and conservation.

A particle beam fusion accelerator. Scientists use the law of conservation of energy to determine what particles are formed by the accelerator. (Reproduced by permission of the U.S. Department of Energy.)

Electricity, heat, nuclear reactions, chemical changes, and other forms of energy are part of everyday life. For each of these applications, some engineer or scientist had to apply the principles of the law of conservation of energy in the design of the machine, device, or system. The electricity we use, for example, is generated by burning coal or oil in power plants or from the running water controlled by hydroelectric dams. Electrical power companies have to know how much fuel to burn or how much water to release over a dam to produce the electricity they need in their systems.

Few concepts have had as much impact on life in the twentieth and twenty-first centuries as the law of conservation of energy. Physical scientists use the law in every calculation that involves energy changes. For example, the law is critical in experiments on subatomic particles. These particles are created when protons, electrons, or other small particles are caused to smash into each other in particle accelerators ("atom smashers"), releasing huge amounts of nuclear energy. Scientists can determine what new particles are formed only when they apply the law of conservation of energy to the reactions that occur in these experiments.

Physical Science

ESSAYS

NINETEENTH CENTURY ADVANCES IN ELECTROMAGNETIC THEORY

Overview

Experiments conducted early in the nineteenth century showed that electricity and magnetism are not separate forces. Investigators found that the flow of an electrical current can produce a magnetic field and a changing magnetic field can generate an electric current. The Scottish physicist **James Clerk Maxwell** (1831–1879; see biography in this chapter) devised a mathematical theory showing how electricity and magnetism are related. That theory eventually had enormous implications both for pure science and for practical applications. An untold number of modern inventions, ranging from radio and television to X-ray therapy and microwave ovens have resulted from Maxwell's work.

Background

Humans have known about and used magnetism and electricity for thousands of years. Early humans must have come across certain rocks in the Earth's crust that were magnetic, and they must have encountered cases in which electrical sparks were produced by the friction between two objects. One of the earliest scientists to write about these forces was the English physicist and physician William Gilbert (sometimes spelled Gylberde; 1544–1603). In 1600, he published *De Magnete* (Concerning Magnets), in

Words to Know

electric current: The movement of electrical charges through some material.

electrical generator: A device for converting magnetic force to electricity.

electromagnetism: A fundamental physical force of nature that can be observed as both electrical and magnetic phenomena.

frequency: The rate at which a wave passes a given point in a certain period of time.

magnetic field: An area of space in which a magnetic force can be detected.

radio wave: A type of electromagnetic wave with a long wavelength and short frequency.

wavelength: The distances between two peaks or two valleys in a wave.

which he summarized all that was known about magnetism at the time. Gilbert also studied and wrote about "electrics." He used this term for materials that developed an electric charge when rubbed.

In the nineteenth century evidence began to surface that electricity and magnetism were related. In 1820, the Danish physicist Hans Christian Oersted (1777–1851) placed a compass near a wire through which an electric current was running. He noticed that the compass needle, which is magnetic with needles pointing to the north and south poles, was deflected. He concluded that the electric current had generated magnetic field around the wire.

A year later, the English chemist and physicist Michael Faraday (1791–1867) observed the reverse process. Faraday found that he could generate an electric current by causing a magnet to spin on its axis. In this experiment, Faraday actually created the first electrical generator or motor.

The connection between electricity and magnetism soon attracted the attention of physicists throughout the Western world. As a result, the nineteenth century is marked by a host of new, independent discoveries about each of these forces and the way they are related to each other. Of all of these discoveries, Maxwell's had the strongest impact on the scientific community.

Physical Science

ESSAYS

Michael Faraday's electrical generator. (Reproduced by permission of Corbis-Bettmann.)

Maxwell's interest was not in carrying out more experiments on electricity and magnetism. Instead, he wanted to find a set of mathematical equations that could quantify and describe the discoveries made by Oersted, Faraday, and other scientists. In 1864, Maxwell published the results of his research. He had found four equations that met his requirements. When the strength of an electric current was substituted into an equation, for example, the strength of the resulting magnetic field could be calculated. The equations accurately represented the discoveries physicists had made in their laboratories.

Maxwell's equations went beyond a description of known facts. They also made predictions about the way that energy such as electricity and magnetism behaves. His equations suggested that electricity and magnetism flow in combination with each other in the form of a wave, somewhat like a water wave. Because the wave always has an electrical component and a magnetic component, it is called an electromagnetic wave.

Maxwell's equations showed that an electromagnetic wave always travels at the same speed, the speed of light. This suggested for the first time that light is a form of electromagnetic energy. Up to this time scientists were still struggling to define the nature of light—what it was made up of, how it traveled. The discovery that light is a form of electromagnet energy suggested to scientists that light, magnetism, and electricity are all just part of an electromagnetic spectrum.

Additional experimentation showed that while the speed of the electromagnetic wave remains the same, other properties of an electromagnetic wave can change. For example, the distance between two crests on an electromagnetic wave (the wavelength) can be very large, very small, or anywhere in between. Also, the number of wave crests that pass a given point in a unit of time, such as one second (the frequency of the wave), can also change. As these factors change, the type of energy carried on the wave changes.

Maxwell's research predicted a whole range of different types of electromagnetic waves. Some of those waves, such as light waves, were already well known. But others had never been seen or imagined. For example, Maxwell's theory suggested the existence of waves with very large wavelengths and very low frequencies. These waves were discovered in **1888** by German physicist Heinrich Hertz (1857–1894). Those waves were named radio waves. Other waves were also predicted with very short wavelengths and very high frequencies. These waves were discovered in **1895** by German physicist **Wilhelm Roentgen** (1845–1923; see biography in the Medicine chapter).

Since the nineteenth century many other forms of electromagnetic radiation have been discovered. These include cosmic rays, gamma rays, ultraviolet radiation, infrared rays, radar, and microwaves. These forms of radiation all differ from each other in their wavelengths and their frequencies.

Impact

Advances in nineteenth-century concepts of electromagnetism moved rapidly from experimental novelties to prominent and practical applications. At the start of the century gas and oil lamps burned in homes, but by the end of the century electric light bulbs illuminated an increasing number of electrified homes.

By mid-century (1865) a telegraph cable connected the United States and England. Yet, within a few decades, even this magnificent technologi-

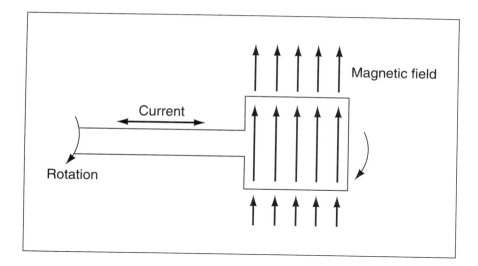

Physical Science

ESSAYS

Illustration of the process of electromagnetic induction. The magnetic field is stationary while the circuit is made to rotate through the magnetic field, creating a current of electricity. (Illustration by Hans & Cassidy. Reproduced by permission of the Gale Group.)

cal achievement was eclipsed by advancements in electromagnetic theory that spurred the discovery of radio waves and the development of the radio that sparked a twentieth-century communications revolution. (See essay "Heinrich Hertz Discovers Radio Waves" in this chapter.) Although the development of radio was largely accomplished in the early years of the twentieth century, its development would have been impossible without the understanding of electromagnetism that was developed in the nineteenth century.

So rapid were the advances in electromagnetism that by the end of the nineteenth century high energy electromagnetic radiation in the form of X rays was used to diagnose injury. In the twentieth century doctors began using gamma rays to treat cancer and other illnesses.

Faraday's method to produce electric current with magnets—known as electromagnetic induction—is a method still used by modern power generators. During the last decades of the nineteenth century, electric motors based on Faraday's electrical generator drove an increasing number of time- and labor-saving machines that ranged from powerful industrial hoists to personal sewing machines. Electric motors proved safer to manage and more productive than steam or fuel burning engines. In turn, the need for the production of electrical power spawned the construction of dynamos, central power stations, and elaborate electrical distribution systems.

Maxwell's equations unifying the experimental work in electromagnetism were more than mathematical interpretations of experimental results. By developing precise formulas with enormous predictive power, Maxwell set the stage for the formation of quantum theory, which suggests that energy transactions occur not only in waves but also in bundles or discrete packets of energy. Maxwell also set the stage for the development of the theory of relativity. Twentieth-century giants such as Max Planck (1858–1947), Albert Einstein (1879–1955), and Niels Bohr (1885–1962) all credited Maxwell with laying the foundations for modern physics. The equations remain a powerful tool for understanding electromagnetic fields and waves. They have many practical applications, including the design of electrical transmission lines and electromagnetic (radio, television, and microwave) antennae.

Simply reciting the types of electromagnetic radiation known today gives one an idea of the number of applications in which they are used. Most forms of communication today make use of some form of electromagnetic radiation, such as radio waves, radar, or microwaves. Each form of radiation in turn has been modified for special purposes, such as AM radio transmission, FM radio transmission, or television sound transmission.

Maxwell developed his four equations for electromagnetic radiation as a way of solving the puzzle about two important forces of nature. It is unlikely that he or his colleagues ever imagined the variety of ways those equations would be put to use in everyday lives of the twenty-first century.

■ FINDING ORDER AMONG THE ELEMENTS: THE PERIODIC TABLE

Overview

The periodic table is one of the most fundamental tools of chemistry. It summarizes information about the chemical elements and shows how they are related to each other. The first widely accepted periodic table was published in 1869 by a chemistry professor named Dmitri Mendeleyev (1834–1907). He began work on his table hoping to help his students learn about the elements. He ended up creating a classification system that helped chemists predict new elements and that led to the discovery of the particles that make up atoms.

Background

The idea of elements first became popular among the ancient Greeks. Greek philosophers argued that the world could not possibly be as complex as it looked. The thousands of different materials we see, they said, must be composed of a handful of basic substances. Various philosophers argued as to what those basic elements were. Some of the most common suggestions were earth, air, fire, water, sulfur, and mercury.

By the late eighteenth century scientists had discovered a number of substances that could not be broken down into any simpler material; they

Words to Know

atomic weight: The weight of an atom expressed in atomic mass units (amu).

chemical element: A substance that can not be reduced to any simpler form of matter.

family of elements: *See* group of elements.

group of elements: A column of elements in the periodic table with properties similar to each other.

law of triads: An early method of classifying elements by groups of three.

periodic table: A chart in which the chemical elements are arranged according to their atomic number and which shows the relationship among the elements.

agreed to call these substances elements. Among the materials that belonged in this category were oxygen, nitrogen, iron, gold, copper, carbon, phosphorus, and sulfur.

In the mid-nineteenth century, about sixty chemical elements were known. One has to say "about sixty" because scientists sometimes disagreed as to whether some materials could really be classified as an element. Some chemists said that those materials could not be broken down into anything simpler. Other chemists claimed that they had been successful in reducing them to simpler materials.

As chemists discovered more and more elements they began to wonder whether the elements were related in any way. Some elements seemed to have similar properties. For example, lithium, sodium and potassium are very similar to each other. They are all soft, light, very active metals. In addition, the properties of certain groups of elements seemed to fall into orderly series. Chemists wondered if there was a way to classify the elements into groups, families, or some other type of arrangement. The arrangement that was eventually discovered is called the periodic table.

Physical Science

ESSAYS

Dimitri Mendeleyev's periodic table with elements arranged by atomic weight. The modern periodic table is organized by increasing number of protons—or atomic number—rather than increasing atomic weight. (Reproduced by permission of The Granger Collection.)

One of the key developments that led to the periodic table was the determination of the atomic weights of the elements. In 1805, English chemist **John Dalton** (1766–1844; see biography in this chapter) stated that every atom of an element has the same weight. This idea implied that it would be possible to measure the atomic weights of the elements. (It is now known that Dalton's hypothesis was incorrect; not all atoms of an element have the same weight. Atomic weights can be measured, but they usually represent an average weight of the atoms in an element.)

Then, in 1809, French chemist **Joseph Louis Gay-Lussac** (1778–1850; see biography in this chapter) proposed that when gases undergo a chemical reaction (such as when hydrogen and oxygen combine to form water molecules) they do so in simple whole number ratios of their volumes. For example, he showed that dinitrogen oxide (N_2O) was formed from two volumes of nitrogen to one volume of oxygen. Jöns Jacob Berzelius (1779–1848), a Swedish scientist, used the ideas of Dalton and

228   Science, Technology, and Society

Gay-Lussac to determine the atomic weights of the sixty-nine elements known at that time. He did so by measuring the relative volume of oxygen with which various elements could combine. He could then infer the atomic weight of the elements from these measurements of volume. He published quite accurate tables of atomic weights in 1818 and 1826.

As a result of Berzelius's work on atomic weights, chemists thought that the properties of elements might be related to their atomic weights. In 1816, the German chemist Johann Wolfgang Döbereiner (1780–1849) noticed a pattern among some elements with similar chemical properties. He noticed that among three elements, chlorine, bromine, and iodine, that the atomic weight of bromine was the average of the weights of chlorine and iodine. Döbereiner called this group a triad. He then found two other triads in which the middle element in each triad had an atomic weight that was the average of the first and third elements. Other chemists tried to find more triads among the elements, but Döbereiner's discovery seemed to be a dead end.

Berzelius was also largely responsible for the fact that the ideas of Amedeo Avogadro (1776–1856) went unnoticed for almost half a century. In 1811, Avogadro had hypothesized that equal volumes of gases contain equal numbers of molecules. If this were the case, the relative atomic weights of gases would be fairly simple to determine. For instance, if one liter of oxygen weighed approximately sixteen times the weight of one liter of hydrogen, then it could be concluded that one atom of oxygen weighs about sixteen times as much as one atom of hydrogen. Avogadro also proposed that two atoms of an element may combine to form one molecule of gas, as in oxygen (O_2) and nitrogen (N_2). It was on this point that Berzelius strongly disagreed with Avogadro. As a result Berzelius and many other scientists of the time wrote incorrect formulas for many compounds. For instance, Berzelius wrote the formula of water as HO rather than H_2O. These incorrect formulas sometimes resulted in incorrect measurements of atomic weight.

In 1858, Italian chemist Stanislao Cannizzaro (1826–1910) showed that the atomic weights of the elements in a molecule could indeed be calculated by applying Avogadro's hypothesis. In addition, he helped to define the difference between atomic weight (that of an atom, such as H) and molecular weight (that of a molecule, such as H_2). Cannizzaro brought forward Avogadro's hypothesis at the first international meeting of chemists. This meeting, called the Karlsruhe Conference, was held in Heidelberg, Germany, in 1860. After Cannizzaro's presentation, Avogadro's hypotheses became widely accepted within a few years.

One of the scientists in attendance at the Karlsruhe Conference was Dmitri Mendeleyev (1834–1907). Mendeleyev was a Russian chemist who happened to be studying in Europe at the time. Mendeleyev made the

Physical Science

ESSAYS

Dmitri Ivanovich Mendeleyev (1834–1907)

Dmitri Mendeleyev was born in 1834 in Tobolsk, Russia. His father was a teacher who died by the time Dmitri was a teenager. When Dmitri was old enough for college, his mother traveled with him across Russia to St. Petersburg, a journey of thousands of kilometers that they took largely on foot. There, in 1850, he enrolled in the Institute of Pedagogy to be educated as a teacher.

After graduating Mendeleyev continued his education at the University of St. Petersburg. He received an advanced degree in chemistry and was awarded a scholarship that allowed him to study in Europe.

After returning to St. Petersburg, Mendeleyev became a chemistry professor. He could not find a textbook suited to his students' needs so he decided to write his own. While working on his book, he began to look for a logical way of arranging the elements. He wrote the name of each element on a note card and listed its properties underneath. Then he began to arrange the cards in different ways, looking for patterns.

Eventually he found that when he ordered the cards by increasing atomic weight, elements with similar properties appeared at regular intervals. In 1871 Mendeleyev published a detailed version of his periodic table. In the

acquaintance of Cannizzaro, from whom he obtained measurements of the atomic weights of elements and a familiarity with the ideas of Avogadro. After returning to Russia, Mendeleyev began searching for a logical way to organize the elements. Eventually, he noticed that when he arranged the elements in order of increasing atomic weight, similar elements appeared at regular, or periodic, intervals. Mendeleyev used his observations to make a table that reflected this pattern.

A similar system was developed almost simultaneously by Lothar Meyer (1830–1895) of Germany. Both systems arrange the chemical element in a table, from left to right and top to bottom, according to their atomic weight. When arranged in this way, elements with similar proper-

papers along with these tables, he tried to show that an element's physical and chemical properties were a function of its atomic weight. He called this relationship the periodic law.

As he was working on his table, Mendeleyev began to suspect that there were elements that had yet to be discovered. He left blanks in his table to accommodate these elements and even made predictions of their properties based on his periodic law. Three of these—gallium, scandium, and germanium—were discovered within twenty years of the publication of Mendeleyev's first table. When the scientific community realized that his predictions were accurate, Mendeleyev became quite well known and was frequently invited to give lectures throughout Europe.

Mendeleyev was not as well respected in Russia. He was considered controversial because he allowed women to attend his lectures and because he openly expressed opposition to the Russian government. Although he was denied admission to the Russian Academy of Sciences, he was made director of the Bureau of Weights and Measures in 1893. Throughout his life he continued to write about chemistry as well as other topics, including education and art. In 1906 Mendeleyev missed being awarded the Nobel Prize in chemistry by a single vote.

ties lie above and below each other in vertical columns now known as groups or families. Elements that lie next to each other in a horizontal row also have similar properties.

Impact

Credit for inventing the periodic table is usually given to Mendeleyev rather than Meyer for one reason: Mendeleyev showed how the periodic table could be put to use by chemists. Mendeleyev's earliest version of the periodic table had some empty spaces. These empty spaces occurred because Mendeleyev insisted on placing elements above and below similar elements. If an element didn't "fall into place" as he drew up the table, be moved it along until he found a place where it fit.

Physical Science

ESSAYS

But then, Mendeleyev did a remarkable thing. He said the reason that empty spaces existed in his table was that some elements had not yet been discovered. Furthermore, he could tell what those elements would be like because of the elements above and below them in the table. Within twenty years of the publication of Mendeleyev's first periodic table, three new elements were discovered to fill empty spaces in the table.

While Mendeleyev's table forms the basis of the modern periodic table, the modern table is organized by increasing number of protons—or atomic number—rather than increasing atomic weight.

Today, the periodic table has two important uses in chemistry. First, it helps chemistry students learn about the elements more easily. One doesn't have to study one hundred individual elements one at a time. Instead, one can learn about a handful of chemical families of elements.

Second, researchers and engineers use the periodic table in their everyday work; for example, imagine that a particular element has long been used to make a type of steel alloy. If researchers would like to make a new type of alloy similar to the old one, one approach is to look at the periodic table. The new alloy might contain an element similar to the one in the old alloy.

Scientists also used the periodic table to help them find a new element to purify water. Chlorine gas is often used to purify the water in swimming pools, but chlorine has some disadvantages when used for this purpose. For example, many people are allergic to chlorine. Chemists looked to the periodic table and determined that bromine, the element beneath chlorine in the periodic table, might be also be effective. Many water purification systems now use bromine rather than chlorine. The discovery made by Dmitri Mendeleyev more than 130 years ago still affects our daily lives in countless ways.

■ LORD RAYLEIGH'S THEORY OF SOUND

Overview

The nineteenth century saw the development of the mathematical techniques and experimental methods needed to understand the vibrations of objects and the motion of sound waves in air and other media. An important contributor to that development was John Strutt (1842–1919), better known as Lord Rayleigh. In 1877, Rayleigh wrote a book entitled *Theory of Sound*. In that book, Rayleigh summarizes most of what was then known about the subject of acoustics (the study of sound). The book also contains reports of many of Rayleigh's own experiments on acoustics. Rayleigh's text remains one of the fundamental reference works for acoustical scientists and engineers.

Words to Know

acoustics: The science dealing with sound.

calculus: A field of mathematics that deals with the rate at which things change.

cochlea: A bony mass in the inner ear that vibrates when struck by sound waves.

elasticity: The ability of a body to return to its original size after being stretched.

seismology: The study of earthquakes.

sonar: A method for detecting and locating objects, especially objects under water, by bouncing sound waves off of them.

ultrasound: Vibrations with properties similar to those of sound, but with frequencies beyond the range of human hearing.

Background

The history of the science of acoustics is somewhat unusual in that, unlike the case of heat or motion, the essential nature of sound has been correctly understood since the time of the ancient Greeks. They identified the origin of sound with the vibration of bodies and understood it to be transmitted through the air in some fashion. The fact that vibrating strings under tension produced harmonious sounds if their lengths were in simple numerical ratios was known to the disciples of Greek mathematician Pythagoras (c. 580–500 B.C.).

Other early scientists also studied the connection between vibrating objects and sound. For example, Italian physicist Galileo Galilei (1564–1642) discussed vibrating bodies in his famous text *Discourses Concerning Two New Sciences*. Similar work was done by the English physicist Robert Hooke (1635–1703). Hooke discovered the law of elasticity, which describes the behavior of a spring. Hooke tried to discover the connection between a vibrating string and the sound it produced.

Physical Science

ESSAYS

Some of the best mathematicians of the seventeenth and eighteenth centuries also worked on the subject of acoustics. They applied the newly developed mathematics of calculus (mathematics related to the rate at which things change) to the solution of such problems. Some of the most ingenious work was done by the French engineer Jean Baptiste Fourier (1768–1830). Fourier was studying the way heat flows through matter. But he noted that the mathematical equations he developed for heat flow could also be applied to the production of sound.

John Strutt, better known as Lord Rayleigh. (Courtesy of the Library of Congress.)

The connection between sound and hearing was studied by the German physicist and physiologist Hermann Helmholtz (1821–1894). He found that the structure through which humans hear sounds is the cochlea in the ear. He demonstrated that vibrations of a sound cause the cochlea to vibrate also. Helmholtz then explained how the quality of musical tones results from their being a combination of different frequencies (the number of wave crests that pass a given point in a unit of time). He summarized his research in his book *On the Sensations of Tone,* first published in 1863.

Lord Rayleigh was strongly influenced by Helmholtz's book. Rayleigh was inspired to begin his own research on the methods by which sound is produced and transmitted. During his life, Rayleigh published 128 papers on acoustics. In 1877, be brought together these papers in *The Theory of Sound*. This book also summarized most of what earlier researchers had learned about acoustics.

Impact

Acoustics play many roles in the lives of humans. Throughout most of history, humans used sound without really understanding how it was produced, transmitted, or received. For example, one does not have to know a mathematical theory of acoustics to hear, to play a violin, or to build a concert hall. Many of the inventions based on acoustics appearing about the time of Rayleigh's book did not depend on knowing the mathematics of acoustics; for example, American inventor **Thomas Edison** (1847–1931; see biography in Technology and Invention chapter) demonstrated the first phonograph in the same year Rayleigh's book appeared, 1877. A year earlier, American inventor **Alexander Graham Bell** (1847–1922; see biography

Physical Science

ESSAYS

The Fessenden oscillator after its first test in 1914. The Fessenden oscillator was the first successful sonic radar, or sonar, system. (Reproduced by permission of the National Oceanic and Atmospheric Administration Central Library.)

in Technology and Invention chapter) had received a patent for the first telephone system. Inventions such as these, however, increased scientists' and inventors' interest in fully understanding the fundamentals of sound production, transmission, and detection.

Scientists would soon discover that they could apply the formulas and principles found in Rayleigh's book to improve the many new sound devices being invented and even to solve acoustical problems in buildings. For example, in 1895 Harvard University built a beautiful new lecture hall with one serious problem. There were so many echoes in the hall that no one could hear the teacher.

Harvard asked American physicist Wallace Sabine (1868–1919) to study this problem. Sabine found a way to take photographs of the sound waves traveling around the room. He then used mathematical equations to calculate the best arrangement of walls, ceiling height, seats, curtains, floor-covering, and other elements in the room. He was able to convert the useless room into an effective lecture hall. Today nearly all new concert and lecture halls are constructed only after a careful mathematical analysis of the room's components has been conducted.

Rayleigh's work has led to countless other applications in the modern world. For example, the form of sound known as ultrasound is used in many areas of research and technology. Ultrasound is a form of sound that can not be heard by the human ear. Yet, the production, transmission, and detection of ultrasound follow all of the same rules that apply to ordinary sound.

One application of ultrasound is known as sonar (sound navigation and ranging), developed during World War I (1914–18). In a sonar device, ultrasound waves are produced under water. They travel through water until they meet some object, such as a fish or a submarine. They bounce off that object and return to their source. The time it takes for the sound to return and the pattern of the sound waves that return carry a great deal of information. They can be used to determine how far away an object is, how fast it is moving and in what direction, and, most importantly, whether it is a fish or a submarine.

Acoustics also has many research applications: the science of seismology depends on the study of sound waves produced during earth movements and earthquakes. When large masses of rock shift deep beneath Earth's surface, they send out vibrations. These vibrations can be analyzed in the same way that ordinary sound waves can be analyzed. Seismologists can determine the region of Earth from which the waves came and many characteristics of the rock movements.

THE DISCOVERY OF RADIOACTIVITY

Overview
Few discoveries in the nineteenth century have had as much impact on science and society as that of radioactivity. Radioactivity is a form of radiation (emission of energy) by which atoms of some elements give off particles and high-energy rays from their nuclei (the positively charged center of the atom). The discovery of radioactivity made it necessary for scientists to think about matter in entirely new ways. From the initial discovery scientists came to understand the inner workings of the atom; and a new science was established, nuclear physics. It also led to the development of new tools and techniques in research, medicine, technology, military science, and many other fields.

Background
One of the most basic laws in science in 1890 was that matter is stable; meaning that scientists believed that matter can not change all by itself. A piece of copper, for example, does not change into a piece of iron unless

Words to Know

polonium: A radioactive element discovered by Marie and Pierre Curie.

radiation: Energy and particles given off by radioactive materials.

radioactivity: The process by which matter breaks apart and gives off radiation.

radium: A radioactive element discovered by Marie and Pierre Curie.

stable matter: Matter that does not break down on its own.

tracer: A material whose presence in a material can be detected because of the radiation it gives off.

X rays: A very powerful form of electromagnetic radiation with very short wavelengths and very high frequencies.

acted on by an outside force. Most investigators had accepted this as a basic truth until 1890; in the 1890s a number of unexpected developments took place that changed scientists understanding of matter. By the end of that decade, the concept of stable matter had been shaken at its very foundation.

French physicist Antoine Henri Becquerel (1852–1908) was the first to experiment with radioactivity—a new form of energy he had discovered quite accidentally. Becquerel was very interested in the discovery of X rays by German physicist **Wilhelm Roentgen** (1845–1923; see biography in the Medicine chapter). X rays are a form of invisible radiation powerful enough to pass right through many forms of matter, including the fleshy tissue of the human body. Although Roentgen discovered these rays he did not know much about them except that they were able to penetrate many solid objects. He chose the name X rays since "x" is the mathematical symbol for the unknown.

Becquerel had been studying certain minerals that are luminescent (glow in the dark) when he also became interested in X rays. He thought that the same mechanism that made certain minerals glow might be what produced invisible X rays. He carried out a number of studies in which

Antoine Henri Becquerel
(1852–1908)

Henri Becquerel had already established himself as a respected French physicist when his discovery of radioactivity catapulted him into the ranks of the world's leading scientists. Although the discovery was unexpected, it was not random. Becquerel's background, experience, and particular circumstances positioned him for this historic event.

Becquerel's father, Alexandre Edmond Becquerel (1820–1891), and his grandfather, Antoine César Becquerel (1788–1878), had each been the physics professor at the Natural History Museum in Paris. Edmond Becquerel was especially interested in phosphorescence. He assembled a large collection of luminescent minerals for the Museum.

Edmond Becquerel's son, Antoine Henri (known as Henri), decided to follow the path that his father and grandfather had chosen. Henri Becquerel entered the Paris Polytechnical School in 1872, where he earned an engineering degree. In 1878 he became his father's assistant at the Natural History Museum. Becquerel was awarded the doctorate in 1888 and was elected to the French Academy of Sciences in 1889. Becquerel became professor at the Museum in 1892, after his father's death.

luminescent minerals were placed in sunlight and then tested them to see if they gave off X rays.

During his research, Becquerel made an interesting and totally unexpected discovery. He found that some of the minerals with which he was working, particularly uranium compounds, gave off radiation similar to X rays even when stored in the dark. No one had ever observed X-ray production in the dark. Becquerel decided that he had discovered an entirely new form of invisible radiation which he named Becquerel rays.

This new form of radiation seemed to be related to the element uranium, since everything Becquerel tested that contained uranium produced the streaks, while the other minerals did not. (Exceptions were later attributed to errors.)

Early in 1896 Becquerel heard that a German physicist had just discovered invisible rays that could pass through opaque objects. Becquerel learned that the invisible rays seemed to come from the phosphorescent screen used in the experiments and he wondered whether other phosphorescent materials might also produce these rays (later called X rays).

Becquerel began testing the specimens in his father's collection of phosphorescent minerals. He soon found that uranium minerals emitted invisible rays, which he believed were a form of light. He published his findings during 1896–1897, naming the rays Becquerel rays.

The scientific community did not seem very interested in his finding so Becquerel turned to his attention to other areas of study. He was drawn back to his rays after **Marie Curie** (1867–1934, see biography in this chapter) and Pierre Curie (1859–1906) used them to trace and identify two new elements, polonium and radium (1898). Around that time Becquerel was recognized as having discovered a new property of matter, which Marie Curie named radioactivity. For his discovery of radioactivity Becquerel shared the 1903 Nobel Prize in physics with Marie and Pierre Curie.

Becquerel published his findings in 1896 and 1897, but most scientists were not very interested, since their journals were being flooded with reports on various kinds of invisible rays. Satisfied that he had established his discovery, Becquerel investigated a different topic for the next year and a half. An engineer in London, Silvanius P. Thompson (1851–1916), had also found in 1896 that uranium gave off invisible rays. After learning that Becquerel had already published this result, Thompson likewise dropped this topic.

Other scientists picked up Becquerel's research where he had left off, most notably Polish-French physicist **Marie Curie** (1867–1934; see biography in this chapter) and her husband Pierre Curie (1859–1906). The Curies decided to search for other elements that might give off invisible

Physical Science

ESSAYS

A scientist wearing protective shielding holds a uranium disk. Henri Becquerel discovered in 1896 that uranium gives of invisible rays, a property later termed radioactivity. (Reproduced by permission of the Department of Energy.)

rays like Becquerel's rays, naming this property *radioactivity*. They already knew from Becquerel's work that radioactivity could be observed only with minerals that contain the element uranium. But the Curies found that pitchblende, a uranium ore, emitted more radiation than uranium itself. They set out to find the source of this more intensive form of radiation.

After many long months of purifying uranium ores, the Curies had an answer to their question. They found that uranium ores contain a chemical element that had never before been seen; this element was even more radioactive than uranium. They named the element polonium, in honor of Marie Curie's homeland, Poland. Six months later, the Curie's found a second new element in the uranium ores. They named this element radium, because of the radioactivity it gave off.

The Curies' discoveries were only the first step in unraveling the mysteries of radioactivity. More than two decades would pass before scientists could fully explain why and how radiation was being emitted by uranium, polonium, radium, and other radioactive elements.

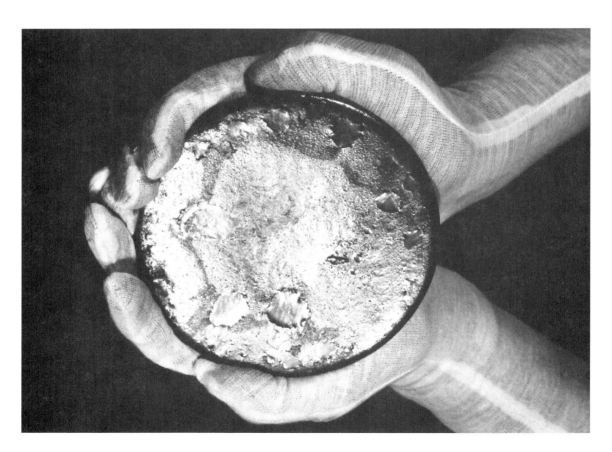

In the late nineteenth and early twentieth centuries scientists realized that the radiation given off by radioactive elements was quite different from the other forms of radiation that they knew about. Investigators identified three types of radiation emitted by radioactive material. One type is emitted as rays, called gamma rays. The other two types are actually particles, called alpha and beta particles.

Scientists found that these subatomic particles are released from the atoms of radioactive materials and when they are released, the radioactive elements change into different elements. Thus the discovery of radioactivity led scientists to the surprising realization that matter is not always stable. As it turned out, some forms of matter do break apart spontaneously, giving off radiation, and changing into new forms of matter. This change can happen very quickly or it can take many, many years.

Physical Science

ESSAYS

Impact

The discovery of radioactivity had an impact on both the physical sciences and the everyday world. Scientists were forced to rethink their assumptions about the nature of matter. It took many years before they were able to fully incorporate this new information into their theories about matter. When they did, a new school of scientific study developed. During the 1930s the field of radioactivity gradually turned into nuclear physics, which later produced the subfield of particle physics.

Radioactivity has a variety of practical applications. Radioactive materials can be used as tracers. A tracer is a material that can be detected because of the radiation it gives off. For example, radioactive fertilizer can be added to a corn field. Investigators can then discover what happens to that fertilizer using instruments that are sensitive to the radiation that is given off by the fertilizer. The path of the fertilizer can be followed as it travels through the stem, leaves, and fruit of the plant. It can also be followed through the soil and groundwater around the plant. Radioactive tracers are also used to locate areas of disease in humans.

Radiation is also a common treatment for cancer. The gamma rays given off by radioactive materials are capable of killing living cells. For this reason, it can be used to treat many forms of cancer. Cancer is a disease that develops when cells begin to grow very rapidly and out of control. These same rays, unless used under controlled conditions, can damage healthy cells. People routinely exposed to radioactive materials must use protective shielding to prevent radiation poisoning.

Advances in nuclear physics raised new ethical, social, and political concerns for the world. Nuclear physicists found a wealth of applications for radioactivity, including the production of nuclear power. Nuclear

Physical Science

ESSAYS

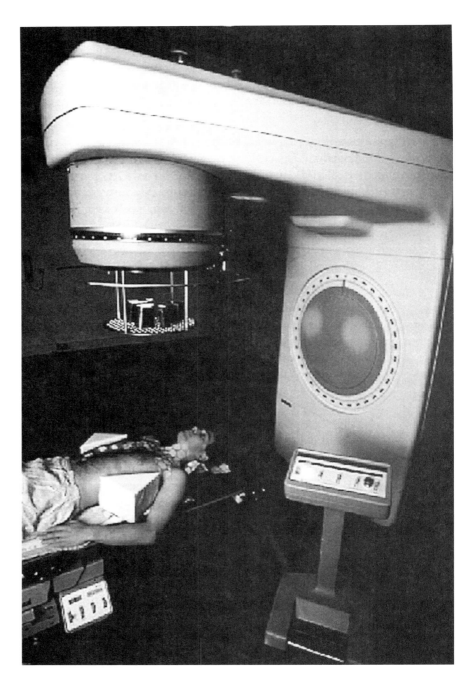

A man undergoing radiation treatment. Radiation is a common treatment for cancer. The gamma rays kill cancer cells. (© Martin Dohrn/Science Photo Library, National Audubon Society Collection/Photo Researchers, Inc. Reproduced by permission.)

physicists also developed terrible new weapons, such as the atomic bomb and the hydrogen bomb. These weapons have far more destructive power than anything previously invented by humans. The creation of such

Words to Know

air: A mixture of nitrogen, oxygen, argon, carbon dioxide, and other gases that make up the Earth's atmosphere; in ancient times, the name used for a gas.

inert gas: A gaseous chemical element that normally does not react with other chemical elements.

neon light: A type of lamp that consists of a glass tube filled with neon or some other inert gas that illuminates when electrified.

oxidation: The process by which a substance reacts with oxygen in the air and changes into a different substance.

shell of electrons: The area in an atom where one or more electrons are found.

weapons has led individuals to question whether it is right to use, or even to create, nuclear weapons.

■ DISCOVERY OF THE INERT GASES

Overview
The last six years of the nineteenth century saw the discovery of five inert gases. An inert gas is a gas that does not react easily with other materials. The fact that these gases are inert does not mean that they have no uses. Indeed, inert gases play some critical roles in our everyday lives today. Inert gases are also called noble gases because they appear "noble"—above interacting with other members of the periodic table.

Background
It took scientists a long time to understand the nature of gases. Part of the problem in working with gases was finding a way to capture them. English physicist and chemist Robert Boyle (1627–1691) was the first scientist to capture a gas in an inverted, submerged flask in 1650. Boyle went on to make an important contribution to scientists' understanding of the behavior of gases. He demonstrated that the amount of space that gas takes up is

Physical Science

ESSAYS

related to the amount of pressure being exerted on it. In the 1770s English chemist Joseph Priestley developed a new method for collecting gases; he captured them over liquid mercury rather than water. Priestley's method led to the discovery of more gases.

During the eighteenth century, all of the most common gases were discovered: carbon dioxide in 1754, hydrogen in 1766, nitrogen in 1772, and oxygen in 1774. By the mid-nineteenth century, then, chemists thought that they had found the major gases that exist on Earth. One puzzle remained for scientists to solve. It is fairly easy to remove oxygen, nitrogen, and carbon dioxide from air. When scientists did this they found that these three gases make up approximately 99 percent of the composition of air. Chemists realized that something else must be present to account for the remaining 1 percent.

In the early 1890s, the English physicist John William Strutt (also known as Lord Rayleigh; 1842–1919) and the Scottish chemist William Ramsay (1852–1916) solved this puzzle. They found a way of trapping and analyzing the last 1 percent of air that had never been identified. They found a new gas in that sample that did not react with other elements. They called the gas argon. Before long, Rayleigh and Ramsay had found four other new, nonreactive gases along with argon in the last 1 percent of air. They called those gases helium, neon, krypton, and xenon. In 1900, the last member of the inert gas family, radon, was found by German physicist Friedrich Ernst Dorn (1848–1916).

Chemists soon began to study the inert gases very carefully. They wondered what it is about these gases that makes them "inert"—why they do not react with other materials the way other chemical do? In the late nineteenth and early twentieth centuries scientists discovered the answer to that question and at the same time made great strides in understanding atomic structure.

The extreme unwillingness of inert gases to undergo chemical reactions eventually led investigators to consider the atomic structure of these elements, which led to the first theories of electron shells. Scientists discovered that atoms have particles called electrons that make up the outer shell of the atom. They theorized that inert gasses have a "full outer shell" of electrons that makes them resistant to reactions with other elements. A "full outer shell" of electrons for an atom is eight electrons. Eventually they proved that all of the inert gases (except helium) have eight electrons in their outer shell. Helium is different because it has only two electrons. For helium, two is a full outer shell.

Scientists found that if atoms have more or less than eight electrons in their outer shell, they react easily with the atoms of other elements to lose or gain electrons to fill up those shells. That attempt to lose or gain electrons is what makes atoms active. If atoms have eight electrons in their outer shell, they tend to resist reactions to other elements. This theory explained why inert gases are so resistant to chemical reactions and also proved very useful to scientists in explaining the behavior of various elements, why chemical reactions take place, and why the periodic table looks like it does.

Impact

At first, because of the difficulty in combining inert gases with other elements, chemists wondered if inert gases would be of any use to humans.

A case at the Library of Congress holds the original Constituion of the United States and the Declaration of Independence. The documents are sealed under glass and surrounded by helium. Inert gases such as helium are useful for preserving documents. **(Reproduced by permission of Archive/Hulton Getty Picture Library.)**

Physical Science

ESSAYS

Neon signs illuminate Piccadilly Circus in London at night during the early twentieth century. "Neon" signs can also contain other inert gases such as argon, xenon, or krypton. The different gases create different colors when electrified. (Reproduced by permission of Archive/Hulton Getty Picture Library.)

Eventually, scientists discovered many uses for these gases because of their low reactivity.

One of their most common uses is to prevent objects from reacting with the elements. For example, the Declaration of Independence and the United States Constitution are very precious documents. Inert gases are used to protect these documents from aging and decay. The documents are stored in glass cases that are filled with argon. The argon prevents oxygen in the air from causing the documents to decay and fall apart.

Inert gases are also used by welders. Welders use very hot flames to join two pieces of metal to each other. But at very hot temperatures, metals usually react with oxygen. Instead of joining together, the metals may simply oxidize, the way iron rusts in the air. So welders often encase the metals on which they are working in a container of argon. The argon prevents the metals from oxidizing while they are being heated.

Neon is well known for its use in neon signs or neon lights. Glass tubes of any shape can be filled with neon or some other inert gas. When an electric current passes through the gas, the gas gives off a colored light. Each of the inert gases gives off a different color when it is electrified. The "neon" signs seen in advertising displays may actually contain neon gas; or they may contain argon, xenon, krypton, or some other gas.

Inert gases taught chemists a lesson. They learned that "not active" does not also mean "not useful," as many uses were discovered for these inert gases.

Physical Science

ESSAYS

■ HEINRICH HERTZ DISCOVERS RADIO WAVES

Overview

In 1888, German physicist Heinrich Hertz (1857–1894) produced and detected electromagnetic waves in his laboratory. With this work, he confirmed the accuracy of **James Clerk Maxwell's** (1831–1879; see biography in this chapter) theory about electromagnetic waves proposed nearly twenty years earlier. Simply producing electromagnetic waves was not sufficient to verify Maxwell's theory. Hertz had also to find a way to detect the waves he produced. Hertz's discovery laid the foundation for countless modern discoveries, including radio, television, radar, telegraph, and microwave communication.

Background

In 1800, the term electromagnetism would have meant nothing to a physicist. Everyone knew about electricity and magnetism, of course. But no one had yet imagined that there might be a connection between these two forms of energy.

Then, in 1820, the Danish physicist Hans Christian Oersted (1777–1851) made an interesting discovery. He found that an electric current generates a magnetic field in its vicinity. Within a short period of time, other physicists discovered related effects. For example, the English physicist and chemist **Michael Faraday** (1791–1867; see biography in this chapter) discovered the opposite effect that Oersted found. Faraday found that a moving magnetic field can produce an electric current.

The next step in understanding the relationship between electricity and magnetism was made in the 1860s by Maxwell. He showed that interactions between these two forces produce what is called electromagnetic radiation in the form of waves in space. He wrote a series of mathematical equations that summarized the main points of his theory. His equations described the nature of those waves and showed that the waves travel at the speed of light, which showed that light is a form of electromagnetic radiation. At the same time, Maxwell's equations suggested that electromagnetic radiation could have either longer or shorter wavelengths which would mean that other types of electromagnetic radiation exist.

Maxwell's theory explained many of the facts then known about electricity and magnetism. Scientists then had to produce evidence that the

Words to Know

electric current: The movement of electrical charges through some material.

electromagnetic waves: Waves that consist of two parts, one electrical and one magnetic.

electromagnetism: A fundamental physical force of nature that can be observed as both electrical and magnetic phenomena.

frequency: The rate at which a wave passes a given point in a certain period of time.

magnetic field: An area of space in which a magnetic force can be detected.

magnetic resonance imaging (MRI): A method for diagnosing disorders of the body by sending radio waves through the body, causing atoms to vibrate in special ways.

optical telescope: A telescope that can be used for observing astronomical objects by means of the light they give off.

oscillation: The regular period motion of an object back and forth between two points.

radio telescope: A telescope that can be used for observing astronomical objects by means of the radio waves they give off.

radio wave: A type of electromagnetic wave with a long wavelength and short frequency.

wireless telegraph: A device for sending messages by means of radio waves transmitted between two stations. Now called a radio.

wavelength: The distances between two peaks or two valleys in a wave.

theory as a whole was correct. One way to produce this evidence was to make predictions from the theory and then test those predictions. If the predictions could be confirmed, the theory was probably correct.

While working with equipment designed to test some properties of electromagnetic fields, Hertz accidentally developed a crude oscillating circuit (an electric current that reversed its direction of flow). In his experiment, Hertz connected a metal wire to a source of electric current. When the current was turned on, it pulsated up and down in the wire at a regular rate. He reasoned from Maxwell's equations that an oscillating electric current in a wire should create an electromagnetic wave. Hertz then set up a similar electric wire at some distance from the wire. He attached a meter to the wire that measures electric current.

Hertz reasoned as follows: When the oscillating current is turned on in the first wire, it should produce an electromagnetic wave. The wave should travel through space and eventually reach the second wire. When it comes into contact with the second wire, it should cause an oscillating electric current to start flowing in that wire. The rate at which the current oscillates in the second wire should be the same as the rate in the first wire.

When Hertz carried out his experiment, he found that his predictions were correct. From Maxwell's equations, Hertz could calculate the properties of the wave produced. He could predict the wavelength and frequency of the wave. The wavelength of a wave is the distance between two consecutive peaks or valleys of the waves. The frequency of a wave is the number of peaks or valleys in the wave that pass a given point every second. Hertz was able to measure an oscillating current in the second wire similar to the one that had flowed through the first wire. Hertz had produced and detected radio waves, which he called "Hertzian waves." He had confirmed Maxwell's theory. More than that, he had demonstrated a principle on which many modern inventions are based.

Physical Science

ESSAYS

Impact
The lasting importance of Hertz's discovery cannot be overstated. Consider the use to which radio and other electromagnetic waves are put today: radio, television, radar, food preparation, welding, heat sealing, magnetic resonance imaging, radio astronomy, and navigation are only a few of the applications.

The most obvious impact of generating and receiving radio waves has been in communications. In 1895, just six years after Hertz's discovery, Italian engineer Guglielmo Marconi (1874–1937) constructed a simple device that used radio waves to ring a bell. By 1901, Marconi had further developed his invention. He was able to send a radio signal from England to Newfoundland. Marconi had discovered the principle of the wireless telegraph and the modern radio. By 1906, voices and music were being sent through the air along radio waves. The era of modern radio broadcasting had arrived, less than twenty years after Hertz's initial success.

Physical Science

ESSAYS

Other inventions followed, including television, communications satellites, and so forth, each simplifying a formerly difficult task—staying in touch over long distances. Prior to radio, rapid communication beyond one's town was difficult and, for most people, rare. Hertz, while not directly involved in changing this, certainly took the first steps by showing it was possible to generate and receive waves that could travel so far so quickly.

Hertz's research had enormous impact in many scientific fields. For example, until 1931 astronomers had only optical telescopes to use in studying the skies. Optical telescopes collect the visible light being given off by stars, planets, and other objects in space. After 1931 scientists discovered that they could also collect radio waves from space; they found that some objects in space that did not give off enough visible light to be picked up by optical telescopes gave off detectable radio waves. Such objects could be detected and studied easily with a receiver tuned to the right frequency—a radio telescope. And today, of course, deep-space probes convey their information and receive instructions via radio signals.

Doctors use radio waves to diagnose illnesses in a technique called magnetic resonance imaging, or MRI. Radio waves are focused on some part of the body. The waves cause molecules in the body to vibrate. The signals sent off by the vibrating molecules provide information about abnormalities in the body.

Many applications of radio waves depend on the ability to use other kinds of electromagnetic waves similar to radio waves. Radio waves are produced when an electric current oscillates in a wire at a certain speed. But speed up or slow down the rate of oscillation, and other kinds of waves, such as radar or microwaves, are produced.

Radar waves, a form of radio-frequency radiation, have been bounced off the moon, Venus, Mercury, and a number of asteroids to learn their distances and to map their surfaces. Radar is also used extensively in weather research, helping to predict and analyze incipient storms. The use of radar for air traffic is well known, of course, as is its use for police speed traps.

Microwave ovens use radio-frequency radiation to cook food, while other microwave devices are used to weld plastics, and seal bags. Radio-frequency radiation is also used for joining metals in some industries.

In summary, the societal impacts of Hertz's research came quickly, were far-reaching, and may be considered ongoing if taking into account the still expanding fields in which radio, radar, and other high-frequency electromagnetic radiation are used. While not as powerful as the development of electronics, it can be argued that radio has transformed more lives than electronics because radios are much more common in the world compared to computers. Virtually every person in the world has reasonable access to radios (excepting, of course, the small percentage of people who live in very remote and primitive areas); the same can hardly be said of computers.

Indeed, the most important indication of the importance of radio to modern society lies in the degree to which we take it for granted. Few question the ability to turn on the radio to hear music or news. Picking up a cordless telephone, a cellular phone, or a walkie-talkie are routine events for most, and we accept as routine that we can see news or sporting events occurring anywhere in the world in real time. Any of these technological commonplaces would have been considered impossible prior to Hertz's discoveries.

Physical Science

ESSAYS

■ CHARLES LYELL POPULARIZES THE PRINCIPLE OF UNIFORMITARIANISM

Overview

Until quite recently, most people believed that the Earth was quite young. Scottish geologist Charles Lyell (1797–1875) was the first scientist to be taken seriously in arguing that the Earth was millions of years old. Lyell's ideas were first presented in 1830 in his work *The Principles of Geology*. The concept that Earth is actually millions of years old dramatically changed the way scientists thought about the world. It also caused people to think about religious beliefs they held about Earth, about humans, and about creation.

Background

Throughout human history, most people have thought of Earth as being quite young and all of Earth's inhabitants as being recently created. In the Western world, the Bible was thought to represent the literal truth about Earth's creation, leading most to conclude that Earth was around six thousand years old. This concept is known as creationism. These beliefs led inevitably to others. A young Earth required direct creation of all plants and animals present today, because there was insufficient time for evolution to work. Creationism does ot allow sufficient time for species to develop, suggesting living things were all created exactly as they are.

The concept of an older Earth was not unknown in scientific circles at the beginning of the nineteenth century. In 1785 Scottish geologist James Hutton had proposed an ancient Earth. Hutton's vision of the history of Earth was of an endless cycle of sediments from the land filling the oceans, turning to rock, and uplifting to form new continents, which then eroded to start the cycle again.

Words to Know

catastrophism: A geological theory that says the Earth as we know it today was produced by a series of sudden and spectacular changes in the Earth's crust in the past by processes that can no longer be observed today.

cosmology: The science dealing with the birth and evolution of the universe.

creationism: A religious belief that the world was created by God out of nothing in a very short period of time, usually as told by one of the two stories in the Biblical book of Genesis.

evolution: The process by which things change over time.

sediments: Solid matter, such as sand and mud, that settles out of a body of water, such as a lake or river.

uniformitarianism: A geological theory that says that the Earth as we know it today has been produced by means of processes that still occur and can still be observed today.

Hutton suggested that the physical changes taking place on Earth have remained the same througout time. There has always been rain, wind, rivers, erosion, earthquakes, volcanoes, and similar processes. The Earth changes very slowly, he said, as a result of processes that we know and understand well. Hutton's theory, later termed uniformitarianism, was not immediately accepted by the scientific community. Most scientists believed in an old Earth too but felt that most features on Earth formed suddenly and catastrophically. The majority of the public, however, believed in a young Earth. These two beliefs—in a young Earth and in Catastrophism—were the generally accepted truths at the beginning of the nineteenth century.

Hutton's ideas regained momentum when they were edited and published by John Playfair in 1802. Playfair elaborated on Hutton's ideas and clarified them in a work titled *Illustrations of the Huttonian Theory*. Playfair also coined a phrase to summarize Hutton's ideas, "the present is the key to the past." This means that the geological processes that can be seen today are the same processes that formed landscapes in the past.

Physical Science

ESSAYS

Tree roots visible because of erosion. Uniformitarianism suggests that processes such as erosion have always occurred as they do now and such ongoing processes are what shape the face of Earth. **(Photograph by Robert J. Huffman. Reproduced by permission of Field Mark Publications.)**

    It was Lyell who, in the 1830s, was most successful in popularizing uniformitarianism. His work, *The Principles of Geology*, explained the principle of uniformitarianism in great detail, pointing to evidence around the world that supported his claim. He made a methodical effort to lay out what was a very controversial theory in meticulous detail, providing over-

Charles Lyell and the Return of the Dinosaurs

Charles Lyell is often called the Father of Modern Geology. One reason he has earned this title is that he made geology into a predictive science. That is, he showed how it was possible to explain changes in Earth's structure by looking at its present state.

Lyell often said that "The present is the key to the past." His point was that we can make intelligent guesses about the past because the same processes are taking place today as occurred throughout history. And those processes can be observed and measured.

However, in one case, he appears to have taken his own ideas a bit too literally. He argued that dinosaurs would once more appear on Earth at some time in the future. They would then replace humans as the dominant species, he said.

Lyell reasoned as follows about the return of the dinosaurs: First, dinosaurs were obviously no longer present on Earth. That meant that something about Earth had changed. He explained this change by assuming that Earth history goes through a series of cycles, somewhat like the changing seasons. Mammals, including humans, were the dominant species during one phase of Earth history, Lyell said. But reptiles, including the dinosaurs, would dominate during another of the Earth's "seasons."

whelming evidence of its truth. By 1850 the idea that the processes that create the landscape occur very slowly and so the earth must in fact be millions of years old was accepted among most geologists.

Impact

Lyell's work had an important effect both on the development of scientific ideas and on the teachings of many religious faiths. The conclusion that Earth and the universe that surround us are the results of long-term, imperceptible change contradicts both biblical and Catastrophic beliefs. This conclusion led other scientists—notably Charles Darwin (1809–1882) and **Alfred Wallace** (1823–1913; see biography in Life Science chapter)—to explore the idea that humans might also be the result of millions of years of imperceptible changes in some other organism. This idea became known as the theory of evolution (see entry on Darwin's theory of evolution in the Life Science chapter). Without the time and the concept of constant, imperceptible change that Lyell's vision gives us, evolution is impossible.

Lyell's estimate of the age of the Earth has had an effect on the way people think about the Earth's creation and the origin of humans. It had affected the way that people view the Bible and the trust that could be placed in it. Many religious people find it difficult to accept both the Biblical version of creation and modern scientific teachings about an ancient Earth. Concluding that Earth is ancient directly contradicts the biblical account of creation. If accepted, one could no longer believe in the Bible as the literal word of God or as an accurate history of Earth and the universe. Instead, one had to believe that the Bible either contained factual mistakes or that it was allegorical. Either of these alternatives carried profound implications for Western religions.

All major scientific theories today accept that Earth is very old. Most experts today set the age at about 4.6 billion years, even older than Lyell's estimate. Modern uniformitarianism differs slightly from it original version. Recent research indicates that catastrophe does play a part in the history of Earth. Meteor impacts are perhaps the best known examples, but large volcanic eruptions, the filling of the Mediterranean Sea, and other such events can significantly and suddenly change the landscape. An important distinction, however, is that the physical processes leading to these catastrophic events remain constant through time.

The subsequent impact of Lyell's work was enormous. Many of the most successful scientific theories of the last 160 years have depended in part on the framework constructed by Lyell. Evolution theory, continental drift, the formation of the universe, and other concepts we take for granted depend on almost imperceptible change over eons (very long periods of time). In astronomy, geology, biology, and other sciences, Lyell's concept of uniformity through time is assumed as a given, with some constraints. And these subsequent theories, in turn, have given us the framework upon which modern scientific knowledge of Earth is based.

Physical Science

ESSAYS

■ DEVELOPMENTS IN ELECTROCHEMISTRY

Overview
Scientists understood well the nature of chemical changes and electricity at the dawn of the nineteenth century, but it was not known what connection there was between these two processes. In 1800, Italian physicist Alessandro Volta (1745–1827) invented the first battery, a device that used

Words to Know

battery: A device that uses chemical changes to generate an electric current.

electric current: A flow of electrons.

electrochemistry: The field of science that involves the relationship between chemical changes and electrical currents.

electrolysis: A process by which an electric current is used to bring about a chemical change.

electroplating: A method by which one metal is laid down on the surface of a second metal by an electric current.

meter: A device for measuring the flow of electric current or some other electrical property.

chemical changes to produce an electric current. Volta's invention inspired other eighteenth century scientists to study the relationship of electric current and chemical changes in more detail. The discoveries they made have a number of important industrial applications today.

Background

In 1771, Italian biologist Luigi Galvani (1737–1798) carried out one of the most famous experiments in physics. He found that if he touched the leg of a dead frog with two different metals, the frog's leg would twitch. Galvani never correctly explained what made the frog's leg twitch. Galvani's work inspired Volta to analyze the experiment in more detail.

Theorizing that the twitching was caused by an electric current created by the combination of the two different metals and fluid in the frog's muscle tissue, Volta decided to repeat Galvani's experiment, but without the frog. He placed two metal wires, one copper and one tin or zinc, into a bowl containing salt water. When he attached a meter between the open ends of the wires, he noticed that an electric current was flowing. A chemical change taking place in the salt water was harnessed by the wires to produce an electric current.

Physical Science

ESSAYS

Electrolysis of water. An electric current causes a chemical change that breaks water down into hydrogen and oxygen. (Photograph by Charles D. Winter. Reproduced by permission of Photo Researchers, Inc.)

Volta then developed a more sophisticated device that worked on the same principle. He made a stack of zinc and copper disks and cardboard circles soaked in salt water. When he attached metal wires at the top and bottom of the stack, he found that a current was produced. Volta had invented the first battery; it was called a Voltaic pile.

Volta had discovered how to produce electricity from a chemical change. In the same year Volta made his first battery, English chemists William Nicholson (1753–1815) and Anthony Carlisle (1768–1840) announced that they had used an electric current to cause a chemical change. They were able to break water down into its component elements, hydrogen and oxygen by applying an electric current to the water. Water is usually a very stable compound of hydrogen and oxygen (H_2O) that does not break down unless some type of energy is added to it (heat or electricity). Nicholson and Carlisle called the process electrolysis. The field of science that involves the relationship between chemical changes and electrical currents is now known as electrochemistry.

Humphry Davy (1778–1829)

Humphry Davy is most famous for his discoveries of potassium, sodium, chlorine, and other elements. Through his research he established the science of electrochemistry. On a more practical note, Davy also invented the miner's safety lamp.

Davy was born to a family of modest means in the remote coastal town of Penzance in Cornwall, England. A mediocre student, he preferred the rocky ocean cliffs to the classroom. At age sixteen he entered an apprenticeship as an apothecary to a local doctor, hoping some day to become a medical doctor himself. At age eighteen he began studying chemistry and performing his own experiments.

The young Davy's chemical research so impressed a few local scientists that in 1798, at age nineteen, he was recommended for a position at Dr. Thomas Beddoes's Pneumatic Institution in Bristol, an institution dedicated to researching the medical uses of gases. It was there that Davy discovered through self-experimentation the intoxicating effects of breathing nitrous oxide.

In 1801 Davy left Bristol for London to accept a position at the recently established Royal Institution. In the next twelve years he enjoyed tremen-

One of the great geniuses of early electrochemistry was English chemist and physicist Humphry Davy (1778–1829). Davy became intrigued by the possibilities of using electrical current to break down compounds that had resisted all other attempts. In 1807, he constructed a very large battery built on Volta's principle. He then passed the electric current produced by that battery through molten (melted) potassium chloride. He found that the compound broke down to form metallic potassium and chlorine gas. Potassium chloride had been known to scientists for thousands of years, but no one had ever been able to break it down into its basic elements.

Over the next year, Davy repeated his experiment with other familiar compounds that had never been decomposed. In the process, he discovered five new elements: sodium, calcium, magnesium, barium, and strontium.

dous success as both a lecturer and a chemical researcher. As a lecturer Davy so thrilled audiences he helped make the subject of chemistry all the rage among London's wealthiest and most fashionable classes.

Davy became most famous, however, for his discovery of potassium and sodium in 1807. He went on to discover numerous other chemical elements and to prove that chlorine, as he called it, is an element rather than an acid. Davy was knighted for his achievements in 1812.

Davy invented the miner's safety lamp in 1815 in response to disastrous explosions in British coal mines. He discovered that by encasing the lamp's flame in wire mesh it would not ignite methane gas trapped in a mine. As reward for his invention Davy was made a baron. Davy served as president from 1820 to 1827 of the Royal Society, Britain's most prestigious scientific institution.

In his final years the ailing Davy traveled, studied natural history, and wrote poetry and metaphysical treatises. Soon after turning fifty he died and was buried in Geneva, Switzerland.

Davy's work was continued and advanced by his most famous pupil, English physicist and chemist **Michael Faraday** (1791–1867; see biography in this chapter). Faraday's greatest achievement was to discover the mathematical laws that explain electrolysis and predict the outcomes of the process. He found out how much of an element could be produced by applying a given amount of current over a given amount of time. Faraday's laws are still used every day by companies who make products by means of electrolysis.

Impact

Electrolysis has become a very important industrial process. It is used primarily for two purposes: for the production of certain chemical elements and for electroplating.

Physical Science

ESSAYS

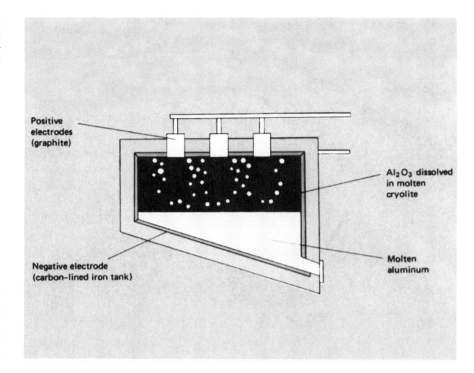

An illustration of Hall-process electrolysis used to make aluminum. (Reproduced by permission of Robert L. Wolke.)

The method for freeing elements from their compounds developed by Davy and Faraday has been improved over the years, but it still is the basic principle by which calcium, chlorine, fluorine, magnesium, sodium, and other elements are obtained. Perhaps the greatest application of this process involves extracting the metal aluminum from its compounds.

Aluminum is a very common element in Earth's crust but it is very difficult to extract. In fact, at one time in the nineteenth century, aluminum metal was very expensive. It cost so much to make that it had no common uses. Then, in 1886 Charles M. Hall (1863–1914), a student at Oberlin College, found a way to use Davy's method of electrolysis to release aluminum from one of its compounds. Almost overnight, aluminum became an inexpensive metal. Today, it is one of the most widely used of all metals.

The methods of electrolysis developed by Davy and Faraday are also used today for electroplating. Electroplating is a method by which one metal is laid down on the surface of a second metal by an electric current. For example, silverware is often made by laying down a thin layer of silver metal on top of a base metal, such as steel. Electroplating is popular because it is a way of protecting the underlying metal from attack by oxygen. It also gives a material a more attractive surface.

Electroplating is done by a method not so different from the one first used by Volta in 1800. One of the wires is placed on the object to be plated. The other wire touches a pure piece of the plating metal. When an electric current flows through the system, the plating metal moves through the solution and attaches to the object being plated. By using Faraday's laws of electrolysis, a person can lay down a layer of metal of exactly the correct thickness on the object.

Physical Science

ESSAYS

■ DISCOVERY OF ICE AGES

Overview

The concept of thick ice sheets covering large portions of the globe is a familiar one. An ice age is a long period of time during which much of the Earth's surface is covered with ice. At the beginning of the nineteenth century, few people could even imagine the concept of an "ice age." But discoveries made and theories invented during that century changed that situation. We know that ice sheets advance and retreat, altering landscape and climate as they do so. The two men largely responsible for making this change were the German-Swiss geologist Johann von Charpentier (1786–1855) and the Swiss-American naturalist **Jean Louis Agassiz** (1807–1873; see biography in this chapter). Charpentier first advanced a reasonable scientific explanation of a recent ice age to explain many of the phenomena found in the Alps, while Agassiz fought to have the idea win acceptance in both the scientific and popular arenas. Acceptance of this theory has changed the way we view the Earth and its climate and affected recent debates on global warming.

Background

The Earth's surface is covered with objects that seem out of place; geologists were long puzzled by certain findings. For example, seashells can sometimes be found on mountain tops, and many areas of northern Europe and North America possess unusual jumbles of sand, gravel, mud, and silt that show no consistent layering. Many of these jumbles contain large rocks that were obviously brought from some other location and left, presumably deposited by some outside agent.

By 1800, scientists, philosophers, and religious teachers had invented a variety of answers for the question of how the jumbles came to be where they are. Some people looked to the Bible for answers. They suggested that the mixture of materials was left behind after Noah's flood. A variation on this theory held that these geological phenomena resulted from icebergs afloat in the post-flood waters, gouging rocks as they floated in shallow waters and depositing rocks and other sediments frozen into their under-

Words to Know

climate: The trend in average weather conditions over a relatively long period of time, usually at least thirty years.

greenhouse effect: The process by which heat is trapped in the Earth's atmosphere by certain chemical compounds, such as carbon dioxide.

ice age: A long period of time in the Earth's history during which much of the Earth's surface was covered with thick sheets of ice.

ice pack: A large, thick mass of ice.

meteorologists: Scientists who study weather, climate, and the atmosphere.

sides. In fact, the term "glacial drift" as a synonym for what is now known as "till" is a carry-over from such theories. Other people suggested that the materials were shot out of the Earth by enormous underground pressures.

By the 1820s, however, a number of geologists had begun to look for more scientific explanations for the origin of these materials. One such person was Jean-Pierre Perraudin, a hunter with an interest in geology. Perraudin convinced Swiss geologist Ignace Venetz (1788–1859) that huge fields of ice had once covered the valleys in the Alps Mountains. The ice had left distinctive marks on the sides of mountains as it moved through the valleys, Perraudin said. And it left behind the sand, gravel, rocks, and other materials when it melted.

Venetz expanded Perraudin's idea over the next several years but did not have much success in convincing the scientific community of its validity. The Swiss geologist, Charpentier, however, was convinced of the accuracy of this theory. Charpentier had greater success in developing the theory to explain the presence of ancient ice flows, but his greatest contribution was passing his ideas along to Agassiz. When Agassiz visited Charpentier in the Alps, the evidence for an "ice age" seemed overwhelming to him. Agassiz became excited about Charpentier's theory and began to add his own ideas to it. He also began to talk, write and teach about the theory in Europe and, later, the United States.

Eventually, the idea of ice ages became more widely accepted. The idea was probably strange for both scientists and non-scientists at first. At first people found it hard to comprehend an ice pack nearly 2 miles (3 kilometers) thick sitting on top of a mountain and that the ice pack could flow down the mountain and through valleys like a giant river. Even more difficult to believe, perhaps, was that such an ice pack once covered much of northern Europe and North America and eventually melted, disappearing completely. But today, almost no one doubts that such events actually did occur. Since the nineteenth century geologists have determined that Earth has experienced many ice ages over the course of more than two billion years.

Physical Science

ESSAYS

Impact

The concept of ice ages has turned out to be a powerful tool for scientists who study the Earth. It explains many of the surface features that had previously been poorly understood. It also turns out to have applications in other fields of science. For example, scientists who study climate now know that weather patterns on Earth over long periods of time are strongly affected by the presence or absence of large ice packs.

Questioning the origins of ice ages has been a fruitful endeavor. Astronomers, geologists, and meteorologists have all developed theories to explain the ice ages. The frequency of northern hemisphere ice advances led to the suggestion by the astronomer Milankovich that regular variations in the inclination of the Earth's axis coupled with regular orbital variations and other factors periodically coincide to lower global temperature long enough to start a glacial advance. Geologists—once the theory of plate tectonics was accepted—suggested that large continents periodically congregate near one of the poles, causing the land to cool and spurring glacial advances. Others feel that long-term variations in weather may follow changes in solar activity, the passage of the solar system through interstellar dust clouds, or other extraterrestrial events. All of these suggestions have resulted in research that, even if inconclusive with respect to ice ages, have led to a better understanding of our climate and factors that may influence it. Perhaps the single most important outcome of all these research efforts, however, is the realization that the Earth's climate does change dramatically over time.

Research efforts at the present time are aimed at determining reasons for sudden climatic shifts, because of fears of global warming caused by human activities. Meteorologists have found that Earth's average annual temperature has been gradually increasing for at least fifty years. Many experts believe that human activities are partially responsible for this change. They suggest that burning coal, oil, and natural gas releases huge amounts of carbon dioxide into the atmosphere. That carbon dioxide, they say, increases

Physical Science

ESSAYS

the amount of heat trapped by Earth's atmosphere. This effect is commonly known as the "greenhouse effect." Scientists studying the environment reason that if we can understand what caused temperatures to change in the past we can better understand whether we can cause global warming.

Scientists believe that the typical temperature of the Earth is much warmer than present global temperatures. In fact, the Earth is currently in what is known as an interglacial period, meaning that the glaciers have temporarily retreated. What scientists do not know is whether the glaciers will advance again or whether the most recent ice age has, in fact, ended, in which case we would expect global temperatures to begin rising, ice caps to begin melting, and glaciers to be retreating. While scientists have not agreed on the time, magnitude, or direction of these changes, they unanimously agree that there will again be a major climatic change. Similarly, whether human activities will hasten an advance, prevent it, or have no impact, cannot be known with current knowledge. By learning more about the ice ages, how they began, what effects they had, and how they came to an end, scientists are trying to understand current environmental changes on Earth.

The concept of ice ages explains many of the surface features of Earth. This lake in Wisconsin was created by a glacier that once covered the area. (Reproduced by permission of JLM Visuals.)

BIOGRAPHIES

■ JEAN LOUIS RODOLPHE AGASSIZ (1807–1873)

Swiss naturalist Jean Louis Agassiz is best known for his research on glaciers that led to the idea of ice ages. An ice age is a period in the distant past when snow and ice covered large areas of the Earth.

Agassiz was born in Motieren-Vuly, Switzerland, on May 28, 1807. He received his education at the universities of Zurich, Switzerland; Heidelberg and Munich, Germany; and Paris, France. He earned a Ph.D. in science in 1829 and a medical degree in 1830.

After graduation, Agassiz moved to Paris to study anatomy with noted scientist Georges Cuvier (1769–1832), who was familiar with Agassiz's early study of fossil fish. The apprenticeship lasted only six months and ended when Cuvier died. In 1832, Agassiz was appointed professor of natural history at the University of Neuchatel in Switzerland. While teaching at Neuchatel, he continued his study of fossil fish. As a result, he wrote a five-volume work on the subject, which listed more than seventeen hundred different species of fossil fish. It is considered Agassiz's masterpiece.

Agassiz was also interested in another subject: glaciers. The mountains in Switzerland are covered with glaciers, so it is not surprising that Agassiz was inspired. During his research, which took place not only in Switzerland, but all over Europe, Agassiz made very interesting discoveries. He observed the surface of glaciers, and especially the effect they left on the Earth.

Jean Louis Agassiz. (Courtesy of the Library of Congress.)

Agassiz continued his experiments and discovered evidence that glaciers move. He found out that a cabin built on a glacier in 1827 had moved nearly a mile down the glacier by 1839. He set up a line of stakes straight across the glacier to test his theory. When he came back later, he found that the stakes had moved down the glacier. In addition, the stakes in the center of the glacier had moved farther than those on the edges. The glacier must have been moving faster in the middle than on the sides.

Physical Science

BIOGRAPHIES

Glaciers in Antarctica. Jean Louis Agassiz studied glaciers in the mountains of Switzerland. He made many interesting discoveries about glaciers and the effect they have on Earth. (Reproduced by permission of JLM Visuals.)

His findings led Agassiz to believe that large masses of ice had once covered very large areas of the Earth at some time in the past. He called that period an ice age. When these large masses of ice melt and shift, they shape geographic regions. At first, Agassiz's theories were met with some skepticism, but eventually they were accepted by the scientific community. (See essay "The Discovery of Ice Ages" in this chapter.) They especially influenced scientists, such as Charles Darwin (1809–1882), who were working on theories of evolution. (See essay "Charles Darwin's Theory of Evolution" in the Life Science chapter.)

In 1846, Agassiz was invited to give a series of lectures in the United States. He was a very good speaker and his lectures were quite popular. Harvard University offered Agassiz a position teaching natural science. He accepted the offer and taught at the university for nearly three decades. He promoted interest in natural history and trained many teachers and researchers in the field. He was also a founding member of the National Academy of Sciences. He is often given credit for making science a more popular subject in the United States.

Agassiz continued his own research in the United States. He found out that a huge glacier had once covered the upper part of the Midwest. When that glacier melted, it formed a large lake now known as Lake Agassiz. The lake dried up and disappeared thousands of years ago.

Agassiz died in Cambridge, Massachusetts, on December 12, 1873. He is buried in Mount Auburn Cemetery in Cambridge, where a glacial rock from Switzerland was used for his headstone.

FRIEDRICH WILHELM BESSEL (1784–1846)

Physical Science

BIOGRAPHIES

Friedrich Bessel was a German astronomer credited with helping establish modern astronomy. Perhaps his most famous accomplishment was measuring the distance to the star known as 61 Cygni. Late in his life, Bessel also studied the orbit of the planet Uranus. From his calculations, he predicted the presence of another planet beyond Uranus. That planet, Neptune, was discovered by the French astronomer Urbain Leverrier (1811–1877) and the English astronomer John Couch Adams (1819–1892) in 1846.

Bessel was born in Minden, Prussia, on July 22, 1784. From an early age he was interested in mathematics and science, but he left school at the age of fourteen to work as an apprentice to an accountant with an import-export company. His interest in the naval activities of the company, combined with his natural interest in science, led him to study the mathematics of navigation, and later astronomy, in his free time. In 1804, he published a paper on the orbit of Halley's Comet. German astronomer Heinrich Olbers (1758–1840) read Bessel's paper and was so impressed that he offered him a job at his observatory. Olbers later claimed that discovering Bessel was his greatest astronomical achievement.

By 1810, Bessel had become quite famous in the field of astronomy. King Frederick William III of Prussia asked him to take charge of the construction of a new observatory at Königsberg. Bessel oversaw the construction and then spent the rest of his life running the observatory. The University of Königsberg wanted Bessel to teach astronomy at the university; since professors were required to have doctorates, the university awarded Bessel a doctorate in 1810 and he remained a professor at the university until his death in 1846.

Friedrich Wilhelm Bessel. (Courtesy of the Library of Congress.)

In the late 1830s, Bessel began to work on the problem of parallax. Parallax refers to the apparent change of position of a distant object when viewed from two different points. Look at an object in the distance first with your right eye closed and then with your left eye closed. The object appears to be in two slightly different positions. The closer the object is to your face, the easier it is to notice this effect. The farther away the object is the more difficult it is to notice.

Physical Science

BIOGRAPHIES

Astronomers had been trying to use this principle for many years. If they could measure the parallax for any astronomical object, they could show definitively that Earth revolves around the Sun instead of the Sun, and everything else in the universe, revolving around Earth. In the sixteenth century Polish mathematician and astronomer Nicholas Copernicus (1473–1543) became a strong proponent of the theory that Earth revolves around the Sun when most scientists believed in an Earth-centered model of the universe. One of the main arguments against Copernicus's model was that, if it were true, one should be able to notice parallax of astronomical objects. If Earth revolves around the Sun, when observers on Earth look at objects in the sky in December and then again in July, they should be seeing the object from two different viewpoints. The position of the star should seem to change between these two times.

By the nineteenth century most scientists accepted that Earth revolves around the Sun, but they could not understand why then they did not perceive any parallax when examining objects in space. It was believed that the stars and other objects in the sky were relatively close to Earth. Scientists thought that the parallax would be fairly easy to measure. Copernicus argued that the stars were much farther away than anyone imagined and so the parallax was very difficult to measure. When parallax was first measured in 1838 it turned out that Copernicus was correct.

Bessel did extensive work in developing and refining methods for accurately measuring the positions of stars. In 1818 he published the positions of more than 3,222 stars. Bessel's careful measurements allowed him to observe very small variations in the positions. He recorded these variations and, by 1838, Bessel figured out the parallax for the star known as 61 Cygni. From the star's parallax, he could estimate its distance from the Earth. He reported that 61 Cygni is about 10 light years (one light year is roughly equal to 5.9 trillion miles) from Earth.

Bessel's work was important for at least two reasons. First, it gave other astronomers a reliable method for estimating the distance to stars. Within a few years, the distance to other stars was being reported by other astronomers.

Second, Bessel's research changed people's understanding of the dimensions of the universe. Prior to his time, most astronomers thought the universe was no more than about one tenth as large as the distance Bessel calculated for 61 Cygni. And 61 Cygni was only one of the closest stars. Suddenly, astronomers realized that they were dealing with a universe much larger than anything they had ever imagined before.

Towards the end of his life, Bessel also became interested in the planet Uranus. Astronomers had known for some time that the planet was not behaving properly. It was not following the orbit that everyone had calculated for it. It seemed that there must be another large object beyond the orbit of Uranus to cause this effect. Bessel did not live long enough to begin a search for the "large object" himself. But Leverrier and Adams used his calculations to search for the planet, which they both found in 1846 and named Neptune.

■ MARIE CURIE (1867–1934)

Marie Curie was a Polish-born French scientist, who was the first woman to earn a Ph.D. in Europe and the first woman to win a Nobel Prize. She later earned a second Nobel Prize, one of only a handful of scientists to be so honored. She is best known for having discovered the radioactive elements polonium and radium and for her research on the chemical and medical applications of radium.

Curie was born Maria Salomea Sklodowska on November 7, 1867, in Warsaw, Poland. Her father was a physics teacher and her mother was the principal of a girls' school. An outstanding student who finished first in her class at the gymnasium (high school), Curie seemed destined to follow in her parents' footsteps. At the time, Warsaw University did ot admit women, and Curie' only chance for continuing her education was to go abroad to study. After working for several years at various jobs, Curie saved enough money to move to Paris, France, in 1891 and enrolled at the Sorbonne, one of the greatest universities in the world.

Her life in Paris was very difficult. She had very little money and often had too little to eat. On one occasion, she even fainted during a class because she was so undernourished. Nevertheless, in 1893 she received a degree in physics, finishing first in her class, and in 1894 she received a degree in mathematics.

While attending the Sorbonne, she met a French chemist by the name of Pierre Curie (1859–1906). They fell in love and were married in 1895. Pierre did not have much more money than Marie, and their wedding present to each other was a pair of bicycles, which they used for their honeymoon trip. After their marriage, they continued to work and study at the Sorbonne. Marie Curie succeeded her husband as head of the physics laboratory at the Sorbonne. Pierre Curie served as professor of general physics in the Faculty of Sciences.

Like many scientists of the time, the Curies were fascinated by the discovery of radioactivity by the French physicist Antoine Henri Becquerel (1852–1908) in 1896. (See essay "The Discovery of Radioactivity" in this

Physical Science

BIOGRAPHIES

Physical Science

BIOGRAPHIES

Marie Curie. (Reproduced by permission of AP/Wide World Photos, Inc.)

chapter.) In fact, it was Marie Curie who invented the term *radioactivity* to describe the phenomenon that Becquerel had first observed.

Inspired by Becquerel, the Curies decided to do further research on radioactive substances, which Marie Curie decided to choose as the subject of her dissertation (a paper submitted for a doctorate degree). They spent more than two years analyzing huge amounts of pitchblende, a naturally radioactive mineral. As a result of their research, they found two new elements: polonium, named for Marie Curie's homeland, Poland, and radium, named for the process of radioactivity. In 1903, the same year Marie Curie received her doctorate, the Curies and Becquerel shared the Nobel Prize for Physics for their study of spontaneous radiation.

The Curies' collaboration ended tragically in 1906 when Pierre was killed in a traffic accident. When Marie Curie was appointed to Pierre Curie's post as professor of physics at the Sorbonne, she became the first woman ever to hold a professorship at the university.

One of Curie's most important achievements was the establishment of a laboratory in her husband's memory. Yielding to Marie Curie and her colleagues' persuasion, the government-funded University of Paris joined the private Pasteur Foundation to fund the Radium Institute.

Marie Curie went on to develop methods to separate radium from radioactive residues in sufficient quantities to study and identify its properties. This would lead to its practical application, especially in medicine. In 1911, Marie Curie would be recognized for this work with a Nobel Prize for Chemistry.

During World War I (1914–18), Curie aided the war effort by assisting in the installation of X-ray machines in ambulances. She was also appointed head of the Radiological Services of the International Red Cross.

Curie continued her research on radioactive materials until the late 1920s. She worked hard to establish a radioactivity laboratory in her native city of Warsaw. Herbert Hoover, president of the United States, presented her with a gift of fifty thousand dollars from the American Friends of Science to purchase radium for the Warsaw laboratory.

By that time, she had developed leukemia, and her health had begun to fail. The leukemia was almost certainly caused by her continuous exposure to radiation. She died at a sanatorium in Haute Savoie on July 4, 1934. The Curies' daughter, Irène, and her husband, Frédérick Joliot-Curie continued their parents' work on radioactivity and were awarded the 1935 Nobel Prize in Chemistry.

Physical Science

BIOGRAPHIES

■ JOHN DALTON (1766–1844)

Englishman John Dalton was a chemist, physicist, and metreorologist who was best known for his work in chemistry. He proposed the atomic theory of matter, one of one of the major steps in creating the modern science of chemistry.

Dalton was born in Eagelsfield, Cumberland, England, in September 1766. His actual birth date is uncertain because his Quaker parents did not register his birth officially. Dalton left school at the age of eleven, and a year later returned to his school as a teacher. He taught young boys and girls who were not much younger than himself.

The field of science in which Dalton first became interested was meteorology, the study of weather. In 1787, he began making daily weather observations with instruments he built himself. He continued to make such observations for the next fifty-seven years, to the day he died. He is

Physical Science

BIOGRAPHIES

said to have collected more than two hundred thousand separate weather observations overall.

Dalton's interest in meteorology eventually led to his study of chemistry and the nature of matter. As he made his observations, he began to think about the composition of air. He decided that air must consist of tiny, individual particles. Dalton went on to say that all forms of matter are made of tiny particles. This idea was not completely new; it had been suggested by Greek philosopher Democritus in the fourth century BC. Democritus called the particles *atomos,* Greek for "unbreakable." Influential Greek philosopher Aristotle disagreed with Democritus, however, and the idea of matter being made of particles fell into disfavor for more than two thousand years.

Dalton revived the particulate theory of matter in the early nineteenth century, calling it atomic theory. He stated that the most basic component of elements are atoms. Elements are substances that can not be reduced to simpler substances. Dalton stated that elements are made up of atoms, with unique atoms for each different element. Lead, for example, is made up of lead atoms, hydrogen is made up of hydrogen atoms, and calcium is made up of calcium atoms. The atoms of these elements all look different.

An important aspect of Dalton's atomic theory was that atoms must have weights which can be determined through experimentation. The first measurements were made by comparing weights of various atoms to that of hydrogen. Hydrogen was chosen as the unit of comparison because it was the lightest substance known. Dalton assigned hydrogen the atomic weight of one; since hydrogen as the lightest substance known, this assured that all atomic weights would be greater than one.

John Dalton. (Reproduced by permission of the Bettmann Archive.)

Scientists knew that hydrogen combines with oxygen to form a molecule of water. Dalton assumed that in this chemical reaction one atom of oxygen combines with one atom of hydrogen. In the laboratory it could be determined that a particular weight of hydrogen combined with eight times its weight in oxygen to form water. It followed then that each atom of oxygen weighed eight times what an atom of hydrogen weighs. Dalton used such experimental methods to create the very first Table of Atomic

Weights. Dalton did not know that ratios of atoms in molecules was not always one-to-one and so his atomic weights were often inaccurate and were adjusted by other chemists throughout the nineteenth century.

Dalton published his atomic theory in 1808 in his book *New System of Chemical Philosophy*. Since it was published, Dalton's atomic theory has guided the design and conduct of a great deal of chemical research. As time went on, his theory was changed, modified, updated, and improved. Today, the atomic theory used by chemists looks little like Dalton's original theory. It still contains the basic idea that particles known as atoms are the basic particles of all forms of matter and that the only way we can really understand matter is to understand the atoms of which it is made.

Dalton made a number of other important contributions in science. He was the first person, for example, to describe color blindness in a scientific publication. He also derived the law of partial pressures. That law states that the pressure exerted by each gas in a mixture is the same as if it were alone in the container.

Dalton was one of the founding members of the British Association for the Advancement of Science in 1831. He died in Manchester on July 27, 1844. It is said that more than forty-thousand people filed past his coffin while he lay in state.

■ MICHAEL FARADAY (1791–1867)

Michael Faraday was an English chemist and physicist who was also a great popularizer of science. He made important discoveries in the fields of cryogenics (low-temperature physics), chemistry, and electromagnetism. He demonstrated electromagnetic rotation (the basis of electrical motors), the laws of electrochemistry, and introduced the concept of electromagnetic fields.

Faraday was born on September 22, 1791 in Newington, Surrey, England. He was one of ten children of a poor blacksmith and his wife. The family was not able to afford a full education for young Michael, and he left school at the age of fourteen. He was then apprenticed to a bookbinder, a job he held for seven years. Faraday was fortunate in finding this job as it gave him a chance to read many kinds of books. He was especially interested in those dealing with science, particularly those on electricity.

At the age of twenty-one, Faraday was offered tickets to one of the regular lectures given by Humphry Davy (1778–1829) at the Royal Institution. Faraday took careful notes of the lecture, which he then illustrated, bound, and sent to Davy's attention. Davy was so impressed with Faraday's work

Physical Science

BIOGRAPHIES

that he offered him a job as an assistant at the Institution. Faraday's first job was washing bottles, although he quickly began to take on more serious responsibilities.

Faraday made his first scientific discovery in 1823 when he developed a method for liquefying gases such as carbon dioxide, chloride, and hydrogen sulfide. He was the first person to find a way of producing temperatures of less than −4°F (−20°C). For this work, Faraday is sometimes called the father of modern cryogenics.

Faraday's greatest achievements centered on the nature of electricity and its effects on chemical changes. In this regard, he took after his master, Davy. Davy had discovered a method for extracting a number of chemical elements from their compounds using an electric current. (See essay "Developments in Eletrochemisty" in this chapter.)

Faraday attempted to follow up on this work by measuring the amount of chemical change produced by a given amount of electric current. He wanted to find out, for example, what weight of hydrogen gas could be released by passing 5 amperes of electricity for one minute through a sample of water. By conducting experiments of this kind, Faraday discovered the first laws of electrochemistry. These laws are still used to predict the weight of an element that can be produced by passing an electric current through a given amount of a compound.

The achievement for which Faraday is probably best known is his work on electromagnetism. In 1820, the Danish physicist Hans Christian Oersted (1777–1851) had found that an electric current can create a magnetic field around it. That discovery was of enormous importance because it showed the connection between two important types of energy, electricity and magnetism. (See essay "Nineteenth Century Advances in Electromagnetic Theory" in this chapter.)

A year after Oersted's discovery, Faraday began his own research on this subject. He soon developed a very primitive motor, in which an electric current passed through a wire caused a second wire to revolve around the first wire. He then explored the idea of reversing this process. He wanted to see if he could use a magnetic field to generate an electric cur-

Michael Faraday. (Courtesy of the Library of Congress.)

274   Science, Technology, and Society

rent. In the early 1830s, he had finally accomplished this task. He devised a system in which a changing magnetic field created an electric current in a wire passing through that field.

Faraday remain affiliated with the Royal Institution for nearly fifty years. To the general public, he was probably best known for the lectures he gave on a great variety of scientific subjects. He believed strongly that everyone should know something about science, and he used his lectures to achieve this goal. Among his most popular lectures was a series given each year at Christmas for young people. One of those series, entitled *The Chemical History of a Candle,* has become one of the great classics on popular science. It has been reprinted many times, and is still available today.

Faraday suffered a mental breakdown from overwork in 1839. Although he recovered, he was never as productive as he had been before his illness. He retired from the Royal Institution in 1862 and died in Hampton Court, Middlesex, England, on 25 August 1867.

Physical Science

BIOGRAPHIES

■ JOSEPH LOUIS GAY-LUSSAC (1778–1850)

Joseph Louis Gay-Lussac was a French scientist who made important discoveries in both chemistry and physics, including groundbreaking work in the study of gases. He developed the law of combing volumes for chemical reactions, and techniques for chemical analysis.

He was born on December 6, 1778 in St. Léonard, Haute Vienne, France. He attended the famous scientific academy, the Ècole Polytechnique in Paris and graduated in 1797. In 1801, Gay-Lussac accepted a position as an assistant to Claude Louis Berthollet (1748–1822). He became a member of a group of talented young scientists who all lived near Berthollet's private laboratory in Arcueil. This group is sometimes called the Arcueil circle, for the town in which the members lived, or the Laplacian school. The latter name comes from the great French physicist **Pierre Simon Laplace** (1749–1827; see biography in Mathematics chapter) who strongly influenced the thinking of group members. Laplace taught that the principles of physics could be used to understand chemical reactions. The combination of two scientific disciplines, called physical chemistry, was somewhat new in the history of science. Several members of the Laplacian school went on to make important contributions to the field of physical chemistry.

In 1802, Gay-Lussac announced his first major scientific discovery. It dealt the effect of heat on the volume of a gas. As the temperature of a gas is increased, Gay-Lussac reported, the gas's volume is increased by the same ratio as the temperature increase. Gay-Lussac found that this law applies to all gases, regardless of their other properties.

Physical Science

BIOGRAPHIES

The French physicist Jacques Charles (1746–1823) had made the same discovery in the 1780s, and the discovery is still best known as Charles' Law. But Charles never published his findings. Since Gay-Lussac did, he deserves at least partial credit for the discovery. In France, for example, the law is usually known as Gay-Lussac's Law.

Gay-Lussac was also very interested in studies of the atmosphere. He made a number of balloon flights, many with his colleague Jean-Baptiste Biot (1774–1862). On one flight, the two researchers reached an altitude of 23,018 feet (7,015 meters). They found that the composition of the air and Earth's magnetic field were no different than at sea level.

By 1806, Gay-Lussac had become a famous scientist in France. Around this time English physicist Humphry Davy (1778–1829) was getting a lot of attention for his work discovering new elements by breaking down compounds using electricity, a process called electrolysis (see essay "Developments in Electrochemistry" in this chapter). The French ruler at the time, Napoleon Bonaparte, wanted to make France the greatest power in the world. He wanted France to be supreme in science, as well as all other areas. He gave Gay-Lussac and a colleague, Louis Thénard, a large grant of money to build a huge battery. Napoleon wanted the scientists to begin discovering new elements with the battery, as Davy was doing in England. As it turned out, Gay-Lussac did not need the battery. He found a new chemical method for breaking down compounds to discover new elements. With this method, he announced the discovery of boron in 1808, nine days before Davy discovered it.

Joseph Louis Gay-Lussac.
(Courtesy of the Library of Congress.)

In 1808, Gay-Lussac reported another important discovery about gases. He had been studying the volumes of gases as they react with each other. For example the gases hydrogen and oxygen combine to form water; and the gases carbon monoxide and oxygen combine to form carbon dioxide. Gay-Lussac discovered that the volumes of the gases that combine in chemical reactions always do so in simple whole-number ratios. For example, two volumes of hydrogen combine with one volume of oxygen to make water; and two volumes of carbon monoxide combine with one volume of oxygen to form carbon dioxide.

The significance of this discovery was later explained by the Italian physicist Amadeo Avogadro (1776–1856). In 1811, Avogadro said that such ratios can be explained by assuming that gases of the same volume have equal number of tiny individual particles, which he called molecules; for example, a volume of hydrogen contains the same number of particles as a volume of oxygen. Along with Dalton's atomic theory, the concept of molecules proposed by Avogadro, based on Gay-Lussac's work, completed the basic principles of modern chemistry.

Later in his life, Gay-Lussac became interested in politics. He was elected to the Chamber of Deputies in 1831 and to the upper house of the French legislature, the Chamber of Peers in 1839. He died in Paris on May 9, 1850.

Physical Science

BIOGRAPHIES

■ JOSIAH WILLARD GIBBS (1839–1903)

J. Willard Gibbs is regarded as one of the greatest American scientists of the nineteenth century and one of the founders of modern physical chemistry. His studies of the physical and chemical processes involving heat and work (work is the ability to exert a force over a distance) developed the science of thermodynamics into one of the most useful tools available to physicists and chemists.

Gibbs was born on February 11, 1839 in New Haven, Connecticut. At the time, his father was professor of sacred literature at Yale University. Gibbs studied science at Yale and, in 1863, received his doctoral degree in engineering. The degree was the first doctorate awarded in engineering in the United States.

After graduation, Gibbs traveled to Europe. He stayed in Europe for three years, attending the lectures of some of Europe's most outstanding mathematicians and physicists. He returned to the United States in 1869 and, two years later, was appointed professor of mathematical physics at Yale. He remained in this post until he died on April 28, 1903.

Early in his life, Gibbs showed some interest in invention. He even applied for and received a patent for a railroad brake in 1866. But he was never very interested in experimentation. He much preferred to think problems through and work them out with paper and pencil. His greatest achievements, then, are in the area of theoretical physical science.

The work for which Gibbs is best known was published in several papers that appeared between 1876 and 1878 in a little-known scientific journal, *Transactions of the Connecticut Academy of Sciences*. In this series of papers, Gibbs showed how the principles of thermodynamics could be

Physical Science

BIOGRAPHIES

applied to chemical reactions. Thermodynamics was a new branch of physics developed by Sadi Carnot (1796–1832), James Joule (1818–1889), Heinrich Helmholtz (1821–1894), and others in the mid-1800s. It deals with the relationship of heat to physical work, especially the transformation of energy from one form to another, such as electric energy to light, or thermal energy (heat) to motion (see essay "The Development of the Concept of Energy" in this chapter). Gibbs had an important insight about thermodynamics. He saw that it could be applied to the atoms and molecules of chemical reactions as well as the boilers and machines that do work.

In these essays Gibbs also developed his theories on phases of matter known as the phase rule. Gibbs related all of the variables involved in a chemical reaction—temperature, pressure, energy, volume, and entropy—and put them together in one equation. Scientists can plug in some variables of the equation and easily determine the other variables. Gibbs's phase rule has also found considerable application in the development and improvement of industrial processes.

Although the importance of Gibbs's accomplishments was immediately recognized in Europe, recognition came more slowly in the United States. His series of papers contained many mathematical equations that showed how the principles of physics could be applied to chemistry. This approach was new to most chemists. For that reason, many chemists did not read or understand Gibbs' reports. But Gibbs remained largely an unknown for another reason. The nineteenth century was a difficult time for anyone interested in the physical sciences in the United States. The study of natural history had long been very popular in this country. Both professional scientists and amateurs searched for rock, plant, and animal samples; developed collections; and wrote descriptions of their discoveries. But there was virtually no scientific establishment in the nation for people who wanted to study chemistry or physics. Indeed, only two men—Gibbs and **Joseph Henry** (1797–1878; see biography in this chapter)—were considered of the caliber of European scientists.

It is largely for this reason that Gibbs' work did not become generally known for more than twenty years. It was finally through the efforts of

Josiah Willard Gibbs. (Courtesy of the Library of Congress.)

Scottish mathematician and physicist **James Clerk Maxwell** (1831–1879; see biography in this chapter) that Gibbs' papers on chemical thermodynamics were eventually translated into both German and French. In 1881 Gibbs received the Rumford medal from the American Academy in Boston and, in 1901, the Copley medal from the Royal Society in England.

■ JOSEPH HENRY (1797–1878)

American physicist Joseph Henry's experiments with electricity and magnetism led to the invention of the telegraph, the electric motor, and the telephone. Henry also served as the first secretary of the Smithsonian Institute in Washington, D.C. In that role, he played a crucial part in the development of American science, insisting that the Smithsonian support original scientific research.

Joseph Henry was born in Albany, New York, on December 17, 1797. His family was very poor and for a time he lived with his grandmother in Galway, New York, working in a general store and doing other odd jobs to help make ends meet. After his father died, when Henry was fourteen years old, he returned to Albany. In 1819, he enrolled in Albany Academy. He taught at county schools and did private tutoring in order to make enough money to remain a student at the academy.

Joseph Henry. (Reproduced by permission of Corbis-Bettmann.)

After graduation Henry worked as a canal surveyor and eventually as an engineer for canal construction. When Albany Academy offered him a position as a professor of mathematics and natural philosophy (physics), he accepted in 1826. He married his cousin Harriet L. Alexander in 1830. During this time he experimented with hot-air balloons, which earned him some recognition and also began his experimentation with electromagnetism and meteorology. In 1832, Henry joined the faculty of the College of New Jersey, now Princeton University.

Henry's primary field of interest was electromagnetic induction. He made electromagnets by wrapping copper wire around a cylindrical bar of iron. When electrical current passed through the copper wire, it made the iron bar magnetic. This principle is still used today to make some of the most powerful magnets used in research and industry.

Physical Science

BIOGRAPHIES

One of Henry's many discoveries was that a system of electromagnets could be used to open and close circuits along a telegraph wire. That principle is used to operate the device known as a telegraph. Henry did not ask for a patent on his invention since he thought all scientific knowledge should be freely available to everyone. Credit for inventing the telegraph usually goes, therefore, to Samuel Morse, who patented the invention. (See essay "Instant Messaging: The Invention of the Telegraph" in the Technology chapter.)

Probably Henry's most important discovery was that of electromagnetic induction. Electromagnetic induction is the process by which an electric current in one wire can create a second electric current in a nearby wire. Henry made this discovery in 1830, but was too busy teaching to write up his results. By the time he was ready to do so, the same discovery had been announced by the English scientist **Michael Faraday** (1791-1867; see biography in this chapter). The law of electromagnetic induction is known today, therefore, as Faraday's Law, not Henry's Law. (See essay "Nineteenth Century Advances in Electromagnetic Theory" in this chapter.) In 1948, a unit of electrical inductance, the henry, was made an international standard in honor of Henry's work in this area.

In 1846, Henry was asked to become the first secretary of the new Smithsonian Institute in Washington, D.C. The Smithsonian Institute had been founded with a large grant of money from a British industrialist named James Smithson. At the time he took his new job, there was virtually no scientific research going on in the United States. Students had to travel to Europe to study chemistry, physics, and other fields of science. After accepting the position he threw his enormous drive and knowledge into establishing the Smithsonian as one of the finest scientific institutions in the world.

■ CAROLINE LUCRETIA HERSCHEL (1750–1848)

Caroline Herschel was the first woman to gain wide recognition in astronomy at a time when such occupations were rarely available to women. She discovered three nebulae and eight comets and compiled several large collections of star and nebulae (masses of interstellar gases or dust) positions.

Herschel was born in Hanover, Germany, on March 16, 1750. Herschel's father was an oboist for, and later bandmaster of, the Hanoverian Foot Guards Band. He taught his four sons music, but Caroline was expected by her mother to remain at home and prepare for marriage, which was customary at the time. Despite her mother's expectatons, Caroline's father did encourage her quest for knowledge. In *The Herschel Chronicle,* Caroline Herschel recalls that her father took her: " . . . on a clear frosty night into the street, to make me acquainted with several of the beautiful constellations, after we had been gazing at a comet which was then visible." At the time, she could not have dreamed of the contributions she would make to the study of comets.

When the French occupied Hanover in 1757, Caroline's brother, William Herschel, emigrated to England. In 1767, her father died, and she decided to join her brother, with whom she was very close, in England. Her mother was opposed to her leaving but was unable to convince Caroline to stay in Germany.

William Herschel had become an orchestra director and organist in Bath, England. When Caroline joined him, she took over responsibility of running his household. William gave her voice lessons, and she became a successful singer. William also undertook teaching her mathematics and English.

William had become interested in astronomy after reaching England. He spent many nights observing the sky and building telescopes. Caroline says of this time in *The Herschel Chronicle*:

> Every leisure moment was eagerly snatched at for resuming some work which was in progress, without taking time or changing dress, and many a lace ruffle . . . was torn or bespattered by molten pitch. . . . I was even obliged to feed him by putting the vitals by bits into his mouth; this was once the case when at the finishing of a 7 foot mirror he had not left his hands from it for 16 hours.

Caroline eventually took up the hobby herself, teaching herself the mathematics needed to calculate star positions.

In 1781, William discovered the planet Uranus. In recognition of this achievement, King George III (1738–1820; ruled 1760–1820) appointed him court astronomer. He and Caroline gave up their musical careers to devote full time to astronomy. Both went on to outstanding careers in the field.

In 1783, Caroline used a telescope given her by William to find three new nebulae in the sky. Three years later, she submitted an important text, *Index to Flamsteed's Observations of the Fixed Stars*, to the Royal Society of London (an organization formed in 1660 where members to met and exchanged scientific ideas). It contained a number of additions (560 stars had been omitted) and

Physical Science

BIOGRAPHIES

Caroline Lucretia Herschel. (Courtesy of the Library of Congress.)

Physical Science

BIOGRAPHIES

corrections to a star catalog prepared by the first Astronomer Royal, John Flamsteed (1646–1719). Her work on this text so impressed the king that he appointed Caroline assistant to her brother in 1787.

William eventually was married in 1788. Although he and Caroline no longer lived together, she continued to perform the mathematical calculations he needed in his work. She also discovered eight comets in 1797. This accomplishment marked the end of her work for many years. Instead, she took over the education of William's son, John. John Herschel (1792–1871) later became a famous astronomer in his own right.

Soon after William died in 1822, Caroline returned to Hanover. There she completed a catalog of twenty-five hundred nebulae in 1828 to assist her nephew in his astronomical work. She received many honors for her work, including the Gold Medal of the Royal Astronomical Society in 1828 and the Large Gold Medal for Science from the King of Prussia in 1846. She died in Hanover on January 9, 1848.

■ CHARLES LYELL (1797–1875)

Scottish geologist Charles Lyell is considered by many to be the "father of geology." In his book, *Principles of Geology*, published in three volumes between 1830 and 1833, Lyell brought together most of the then-known facts about Earth. This work laid the foundation for the modern science of geology.

Lyell was born in Kinnordy, Scotland, on November 14, 1797. He was the eldest of ten children. His father was a naturalist who had a

Charles Lyell.
(Courtesy of the Library of Congress.)

large library, including many books on geology. This may have sparked the young Lyell's interest. At Oxford University, Lyell studied mathematics, law, and geology. He earned his degree in law, but left the field soon after graduation, being much more interested in geology.

Lyell's primary interest was Earth history. He agreed with most scientists of the time that the earth was ancient, although the public largely thought the Earth was relatively young believing that the Earth was formed as described in the Bible. He disagreed, though, with fellow geologists' theory called catastrophism. This theory held that catastrophes, such as the biblical flood, were responsible for the Earth's features.

Lyell found himself more in agreement with another theory introduced by the great Scottish geologist James Hutton (1726–1797)—the theory of the uniformity of causes. According to this theory, continuous geological processes, occurring very gradually over very long periods of time, have shaped the Earth's surface. Lyell expanded and refined this theory (known today as Uniformitarianism) which became the centerpiece of his *Principles of Geology*. (See essay "Charles Lyell Popularizes the Principle of Uniformitarianism" in this chapter.)

Many scientists refused to accept Lyell's teachings about Earth history. One point that troubled them was that such teachings usually meant that plants and animals had evolved over very long periods of time. And this belief was largely contrary to religious teachings in the nineteenth century.

Nevertheless, Lyell's work did inspire the great English naturalist Charles Darwin (1809–1882), who developed the theory of evolution (see essay "Charles Darwin's Theory of Evolution" in the Life Science chapter). Darwin was very interested in geology himself and later applied many of Lyell's ideas in his own book about evolution, *On the Origin of Species*.

The relationship between Lyell and Darwin worked both ways. When Lyell read Darwin's book, he was immediately convinced that Darwin's assumptions about the evolution of plants and animals were true. Lyell went on to explore the question as to what the theory of evolution meant to humans. In 1863, he wrote *The Antiquity of Man*. In this book, he argued that the principles of evolution developed by Darwin must also apply to the human species. Lyell's book preceded Darwin's own book on the subject, *The Descent of Man*, by eight years.

Lyell was knighted in 1848 and made a baron in 1864. He died in London in February 1875. At the time, he was working on the twelfth edition of *Principles of Geology*.

■ JAMES CLERK MAXWELL (1831–1879)

Maxwell was a Scottish mathematician and physicist who devised the modern theory of electromagnetism. Electromagnetism is a force that results from the interaction of electrical and magnetic fields. Maxwell also made important contributions to the fields of astronomy, heat, and molecular kinetics (the movement of gas molecules).

James Clerk Maxwell was born in Edinburgh, Scotland, on November 13, 1831. His family name had originally been Clerk. His father added the name Maxwell after he inherited an estate left to him by a family of that name.

Physical Science

BIOGRAPHIES

Maxwell entered the University of Edinburgh at the age of sixteen. Three years later, he transferred to Cambridge University in England. There he helped form a group that became known as the Cambridge School. Members of the group were especially interested in finding ways to use mathematics to describe physical problems.

Maxwell's earliest accomplishments were in the field of astronomy. In 1857, he proved that Saturn's rings must be made of countless tiny particles, not solid or liquid disks as was then believed. About three years later, he became interested in the velocities (speeds and direction) of molecules that make up gases. He devised a mathematical theory explaining the energy those molecules would have in gases of various temperatures.

In 1856, Maxwell was appointed professor of natural philosophy at Marischal College in Aberdeen, Scotland. Four years later, he took a similar position at King's College in London. In 1865, Maxwell decided that he would leave academic life. He retired from teaching and settled down at the family estate. It was during the years from 1865 to 1870 that Maxwell made his most famous contribution to science: the mathematical theory of electromagnetism.

Humans had long known about both electricity and magnetism. But it was not until the nineteenth century that a connection between these two forces became apparent. In 1820, the Danish physicist Hans Christian Oersted (1777–1851) found that an electric current produces a magnetic field in its vicinity. And a decade later, the English chemist and physicist Michael Faraday (1791–1867; see biography in this chapter) showed that a magnetic field can produce an electric current.

James Clerk Maxwell.
(Courtesy of the Library of Congress.)

The task Maxwell set for himself was to describe these two events with mathematical equations. He was not only successful with this problem, but he also made an additional discovery. In solving the equations he devised, he found that a constant number kept reappearing. That constant number, which he called c, turned out to be the velocity of light. Maxwell had proved that light itself is a form of electromagnetic radiation. In his work, Maxwell had demonstrated the connection among three apparently very different phenomena: electricity, magnetism, and light. (See essay "Nineteenth Century Advances in Electromagnetic Theory" in this chapter.)

In 1871, Maxwell was convinced to come out of retirement. He accepted an appointment as professor of experimental physics at Cambridge, the first person to hold that post. He was not a very good lecturer, and his classes often had no more than a handful of students. Only a few of these students were actually able to understand what Maxwell was saying in his lectures.

Maxwell remained at Cambridge for only eight years. In 1879, he developed colon cancer, the same disease that had killed his mother. He died in Cambridge on November 5, 1879.

BRIEF BIOGRAPHIES

◢ JOHN COUCH ADAMS (1819–1892)

Adams was co-discoverer of the planet Neptune in 1846. He and French astronomer Urbain Leverrier were both puzzled by the irregular orbit of Uranus. They assumed that another planet beyond Uranus was causing the irregular orbit. They calculated where the new planet should be found.

◢ ANDRÉ MARIE AMPÈRE (1775–1836)

Ampère was a French mathematician and physicist who made important discoveries in the fields of electricity and magnetism. He found that an electric current can produce a magnetic field. The unit used to measure electric current in the metric system, the ampere, is named after him.

◢ SVANTE AUGUST ARRHENIUS (1859–1927)

Arrhenius was a Swedish chemist who studied the changes that take place when various substances are dissolved in water. He showed that the molecules of such substances often break apart into two pieces, which he called ions. For his work, Arrhenius was awarded the 1903 Nobel Prize in chemistry.

◢ CARL AUER, BARON FREIHERR VON WELSBACH (1858–1929)

Von Welsbach was an Austrian chemist who invented a new type of gas lantern. The mantle in this lantern is made of cloth soaked in a solution of cerium and thorium nitrates. When heated with a gas flame, this mantle gives off a brilliant white light. This type of lamp is still widely used today.

Physical Science

BRIEF BIOGRAPHIES

▲ HERTHA MARKS AYRTON (1854–1923)

Ayrton was a British electrical engineer who did research on the electric arc. The electric arc uses an electric current to produce a bright light. Arc lamps are used in indoor lighting systems, searchlights, and movie projectors. Ayrton was the first woman elected to the Institution of Electrical Engineers and the first woman permitted to read her paper in person before the Royal Society of London.

▲ JÖNS JACOB BERZELIUS (1779–1848)

Swedish chemist Jöns Jacob Berzelius invented the modern system of chemical notation, and discovered the chemical elements of cerium, selenium, and thorium. His most important theoretical contribution was his theory of electrochemical dualism, a theory that could not stand up to the development of organic chemistry.

▲ GEORGE PHILLIPS BOND (1825–1865) AND WILLIAM CRANCH BOND (1789–1859)

The Bonds were American astronomers who discovered the eighth satellite (moon) of Saturn. The father-and-son team also produced some of the earliest good photographs of the moon, stars, and other astronomical objects. William Bond was appointed the first director of the Harvard Observatory in 1839. Upon his death in 1859, he was succeeded by his son in that position.

▲ ROBERT WILHELM BUNSEN (1811–1899)

Bunsen was a German chemist who made important contributions to the field of spectroscopy. In spectroscopy, the composition of materials is studied by means of the flames they produce when heated. Using this technique, Bunsen was co-discoverer of two elements, cesium and rubidium. He is probably best known for the improvements he made in laboratory burners, which now carry his name, bunsen burners.

▲ JOHANN VON CHARPENTIER (1786–1855)

Charpentier was a German-Swiss geologist who studied ice ages in Europe. He built on the work of Swiss geologist Ignatz Venetez (1788–1859). One of his most important contributions was to interest the Swiss-American naturalist Louis Agassiz (1807–1873) that ice ages had actually occurred in the past.

▲ RUDOLF JULIUS EMMANUEL CLAUSIUS (1822–1888)

Clausius was one of the discoverers of the second law of thermodynamics. He also helped found the branch of science known as thermodynamics.

Clausius also helped develop the concept of entropy. Entropy is the amount of disorder in a system. It can be used to determine the amount of energy that can be converted to useful work in the system.

◮ WILLIAM MORRIS DAVIS (1850–1934)

Davis was an American geologist who made advances in many fields. Probably his most important contribution was showing how erosion changed the shape of landforms over very long periods of time. He also studied the origins of limestone caves and coral reefs and islands. His textbook, *Elementary Meteorology,* was one of the most widely used books in the field for more than thirty years.

◮ JAMES DEWAR (1842–1923)

Dewar was a Scottish chemist and physicist who invented a container for keeping hot liquids hot and cold liquids cold. The Dewar flask is also known today as a thermos bottle. Dewar used his invention to study very cold, liquefied gases. During his research, Dewar became the first person to liquefy hydrogen gas and to discover the magnetic properties of liquid oxygen.

◮ JOHANN FRANZ ENCKE (1791–1865)

Encke was a German astronomer best known for his research on comets. He carefully studied the orbit of a comet that was later given his name. He found that the comet traveled around the sun in a very short period, only 3.3 years. He used this information to write a mathematical formula for the orbit of the comet. He was later able to use that formula for the orbits of other comets.

◮ ARMAND-HIPPOLYTE-LOUIS FIZEAU (1819–1896)

Fizeau was a French physicist who, in 1849, made the first accurate measurement of the speed of light. He later found a way of measuring the speed with which stars are moving toward or away from Earth. The methods developed by Fizeau are still widely used by astronomers in measuring the velocity of galaxies, the motion of gases around black holes, the expansion of the universe, and other astronomical phenomena.

Physical Science

BRIEF BIOGRAPHIES

James Dewar. (Courtesy of the Library of Congress.)

Physical Science

BRIEF BIOGRAPHIES

▲ JEAN-BERNARD-LÉON FOUCAULT (1819–1868)

Foucault was a French physicist who demonstrated Earth's rotation using a pendulum. The Foucault pendulum consists of a long wire holding a heavy ball. As the pendulum swings back and forth, Earth (and the floor beneath the pendulum) rotates beneath it. An observer can see how the alignment of pendulum and floor change over time. Foucault also invented the gyroscope, studied the composition of crystals, and developed a better method for making telescope mirrors.

▲ AUGUSTIN-JEAN FRESNEL (1788–1827)

Fresnel was trained as a civil engineer and worked most of his life in that field. However, he was also very interested in the study of optics and spent most of his free time working on that subject. He became convinced that light travels in waves, like water waves. He was unaware that a similar theory had been developed by Dutch physicist Christiaan Huygens (1629–1695) and English physicist Thomas Young (1773–1829). Fresnel was able to show that the diffraction of light could be explained by the wave theory.

▲ HERMANN HELMHOLTZ (1821–1894)

Helmholtz was a German physicist, philosopher, and physiologist. He invented an instrument for studying the human eye, and performed a great deal of research concerning the inner ear and nerve impulses. Helmholtz also studied the transformation of energy of from one type into another, which was important to the popular concept of a mechanical and dynamic theory of heat.

▲ WILLIAM HENRY (1774–1836)

Henry was an English experimental chemist who helped prove the validity of John Dalton's atomic theory. He formulated Henry's Law, which describes the relationship between mass and pressure for a gas dissolved in liquid. Henry also wrote the most influential chemistry textbook of his time, which stood as the standard for over thirty years.

▲ MARGARET LINDSAY HUGGINS (1848–1915) AND WILLIAM HUGGINS (1824–1910)

The Huggins were a husband-and-wife team of Irish-English astronomers. They used the methods of spectroscopy to show that the elements that make up the stars are the same as those found on Earth and in the Sun. They also discovered that nebulae are made of gases and they found that comets contain hydrocarbons, compounds of hydrogen and carbon. The Huggins also made many measurements on the velocities of stars moving away from the Earth.

▲ JAMES PRESCOTT JOULE (1818–1889)

Joule was a British physicist who studied the relationship of mechanical, electrical, and heat energy. He discovered the connection between heat and electricity, a relationship now known as Joule's law. Joule also carried out experiments that helped proved the principle of conservation of energy. The SI unit of energy, the joule, is named after him.

▲ GUSTAV ROBERT KIRCHOFF (1824–1887)

Kirchoff was a German physicist who made fundamental contributions to the understanding of electricity and light. He developed the concept of a "black body," an object that gives off light whose color depends only on its temperature. Black bodies have had an important role in the development of many theories in physics. For example, the modern theory of quantum mechanics is based on Kirchoff's original idea about black bodies.

▲ WILLIAM EDMOND LOGAN (1798–1875)

Logan was one of Canada's most famous geologists. He was the first director of the Geological Survey of Canada, serving from 1842 to 1869. He was born and grew up in Wales, where he studied coal deposits. In Canada, he made extensive studies of Precambrian and Paleozoic rocks. His efforts at the Geological Survey insured its success as a major governmental agency.

▲ ERNST MACH (1838–1916)

Mach was an Austrian physicist especially interested in the topic of sound. He discovered that the way air flows over an object changes dramatically as it approaches the speed of sound. Today, scientists often refer to the speed of sound as "mach 1," which is equal to the speed of sound, or some multiple of that number. For example, "mach 2" is twice the speed of sound, "mach 3," three times the speed of sound. Mach is perhaps even more famous for his studies in the history and philosophy of science.

▲ ALBERT ABRAHAM MICHELSON (1852–1931)

American physicist Michelson devoted most of his study to the accurate measure of the speed of light. In 1882 he calculated the most accurate measurement of the speed of light to date, until his own revision nearly

Physical Science

BRIEF BIOGRAPHIES

thirty years later. Michelson was awarded the Nobel Prize in physics in 1907. He was the first American to receive this honor.

▲ MARIA MITCHELL (1818–1889)

Mitchell was the first professional female astronomer in the United States. She grew up helping her father make astronomical observations for New England whaling ships. In 1847 she discovered a new comet. From 1859 to 1868, she worked at the U.S. Nautical Almanac Office. There she spent much of her time working on the orbit and appearances of Venus. In 1868, she resigned her post to become professor of astronomy at Vassar College. She was the first woman to be elected to the American Academy of Arts and Sciences.

▲ WILLIAM NICOL (1768–1851)

Nicol was a British geologist whose fame rests on his invention of the Nicol prism. A Nicol prism is a mineral that converts unpolarized light into polarized light. Unpolarized light is light that vibrates in every direction. Polarized light vibrates in only certain specific directions. Nicol prisms are very useful in the study of rocks and minerals.

▲ HANS CHRISTIAN OERSTED (1777–1851)

Oersted's contributions to science were successful springboards for major discoveries by many other scientists. His demonstrations of the connection between electric and magnetic forces proved that the lines of a magnetic field run in circles around a wire carrying an electrical current.

*Maria Mitchell.
(Courtesy of the Library of Congress.)*

▲ GEORG SIMON OHM (1787–1854)

Ohm was a German physicist who discovered an electrical law that is now named in his honor. Ohm's Law says that the current flowing through a conductor is directly proportional to the potential difference and inversely proportional to the resistance of the conductor. That is, the greater the potential difference, the greater the current, and the greater the resistance, the less the current.

Physical Science

BRIEF BIOGRAPHIES

▲ HEINRICH WILHELM MATTÄUS OLBERS (1758–1840)

Olbers was a German astronomer and physician who made many important astronomical discoveries. He found the second and third known asteroids, Pallas in 1802 and Vesta in 1807. He also developed a method for calculating the orbits of comets and suggested a theory to explain the fact that a comet's tail always points away from the Sun. He also stated Olbers's paradox. That paradox was based on the way astronomers of the time viewed the Universe. They thought the Universe was infinitely large. It should, therefore, contain an infinite number of stars. In that case, the sky should be very bright. But it is actually almost entirely dark. That condition is known as Olbers's paradox.

▲ CHRISTIAN FRIEDRICH SCHÖNBEIN (1799–1868)

Schonbein was a German chemist who discovered the presence of ozone in Earth's atmosphere. At the time of Schonbein's discovery, ozone was regarded as a laboratory curiosity. As a result of Schonbein's work, scientists have come to realize the important role that ozone plays in controlling Earth's climate.

▲ BENJAMIN SILLIMAN (1779–1864)

Silliman was an American chemist and geologist who played an important role in establishing a scientific community in the United States. He founded an important publication, the *American Journal of Science and Arts* in 1818. He also helped to make science more popular with the general public.

▲ WILLIAM SMITH (1769–1839)

English mineral surveyor and geologist William Smith recognized that the rock strata, the vertical changes in layers of rock, could be distinguished from one another by the fossils that were held in the rock. He published the first detailed geologic survey of England.

▲ ERNEST SOLVAY (1838–1922)

Solvay was a Belgian chemist who found an inexpensive way to make sodium carbonate, also known as soda or washing soda. Soda is one of the most important of all industrial chemicals. The method he developed is now known as the Solvay process. Solvay was also a very clever businessman and made a fortune on his invention. He used some of his wealth to establish and promote conferences, institutions, and meetings for the advancement of science.

Physical Science

BRIEF BIOGRAPHIES

▲ EDUARD SUESS (1831–1914)

Suess was an Austrian geologist who made many discoveries and developed some important theories about Earth's structure. He suggested that moving land masses cause earthquakes. He also predicted that sea levels had changed over time. He studied fossils from South America, Africa, Australia, and India, and came to the conclusion that these areas were once connected to each other. He called that large land mass Gondwana, a name that is retained today.

▲ JOSEPH JOHN THOMSON (1856–1940)

English physicist J. J. Thomson is best known for discovering the electron. Electrons are the negatively charged particles that make up the outer shell of atoms. Thomson received the 1906 Nobel Prize in physics for his discovery.

▲ WILLIAM THOMSON (LORD KELVIN; 1824–1907)

Thomson was an English physicist whose most influential discovery was the Kelvin scale of absolute temperature. The Kelvin scale places absolute zero, the temperature at which molecules stop moving, at -273°C. The concept of an absolute zero was very useful in the creation and verification of many thermodynamic theories.

▲ JOHN TYNDALL (1820–1893)

Tyndall was an Irish physicist who described the scattering of light by tiny particles. That discovery is now known as the Tyndall effect. One consequence of Tyndall's discovery was that he decided that ordinary air contains microorganism which can cause disease. He placed some food in air that did not show the Tyndall effect. That food did not decay. This discovery was an early step in the development of methods for the preservation of food.

▲ JOHN JAMES WATERSTON (1811–1883)

Waterston was an English physicist whose great talents were largely unappreciated during his lifetime. His most important discovery was the kinetic theory of gases. That theory says that the temperature of a material is a function of how rapidly its molecules are moving. Waterston reported his theory to the Royal Society, which decided not to publish it. It was not until after his death that Waterston's genius was fully appreciated.

▲ WILLIAM HYDE WOLLASTON (1766–1828)

Wollaston was an English scientist who made contributions in astronomy, biology, chemistry, medicine, and physics. He discovered the elements palladium (in 1804) and rhodium (1805). He also developed a method for purifying a number of metals. He studied the dark lines in the Sun's spectrum and invented the camera lucida, a primitive method of photography.

▲ THOMAS YOUNG (1773–1829)

Young was an English physician and physicist who revived the theory the light travels in waves as opposed to particles. He is best remembered for his double-slit experiment demonstrating the interference of light, an absolute measure for the elasticity of solids known as Young's modulus, his optical studies, and his contributions to deciphering the Rosetta Stone.

RESEARCH AND ACTIVITY IDEAS

- The first batteries developed by Alessandro Volta were very simple. Find out how these batteries were constructed. Then construct a simple battery yourself to show how it works. Write a short statement telling how your model battery is different from other batteries in use today.

- Scientists in the nineteenth century learned that work and heat are closely related to each other. Read about (1) ways in which heat is used to do work and (2) ways in which work results in the production of heat in today's world. Select any one example from each category and make a drawing or diagram that illustrates the principle involved.

- The English geologist William Smith found a way of estimating the age of rocks by means of the fossils found in the rocks. Read about the system he invented. Then use modeling clay to illustrate how the system worked. Finally, write a short statement explaining how Smith's research affected the religious beliefs of people who lived in his day and of people who live today.

- The greatest achievements in astronomy during the nineteenth century included finding out how to measure distances to the stars, determining the composition and temperature of stars, and calculating the speed at which stars move through space. Some people have argued that this research, while interesting, has no practical value. Imagine that you are an editorial writer for your local newspaper. Write an article that takes one of two positions: (1) research in astronomy is important to the everyday lives of people and should be funded at least in part by tax dollars, or (2) research in astronomy has no practical value for ordinary people and should not be funded with public tax dollars.

FOR MORE INFORMATION

Books

Blundell, D., and A. C. Scott, eds. *Lyell: The Past is the Key to the Present.* London: Geological Society (Geological Society Special Publication No. 143), 1998.

Campbell, Lewis, and William Garnett. *Life of James Clerk Maxwell.* New York: Johnson Reprint Corporation, 1970.

Cantor, Geoffrey, David Gooding, and Frank James. *Michael Faraday.* Boston: Humanity Books, 1996.

Cobb, Cathy, and Harold Goldwhite. *Creations of Fire: Chemistry's Lively History from Alchemy to the Atomic Age.* New York: Plenum Press, 1995.

Coulson, Thomas. *Joseph Henry: His Life and Work.* Princeton, NJ: Princeton University Press, 1950.

Crosland, Maurice. *Gay-Lussac: Scientist and Bourgeois.* Cambridge: Cambridge University Press, 1978.

Goldman, Martin. *The Demon in the Aether: The Story of James Clerk Maxwell.* Edinburgh: Paul Harris Publishing, 1983.

Gould, Stephen J. *Time's Arrow, Time's Circle.* Cambridge, MA: Harvard University Press, 1988.

Hallam, A. *Great Geological Controversies.* Oxford: Oxford Science Publications, 1989.

Hunt, Frederick V. *Origins in Acoustics: The Science of Sound from Antiquity to Newton.* New Haven, CT: Yale University Press, 1978.

Levinson, Thomas. *Measure for Measure: A Musical History of Science.* New York: Simon & Schuster, 1994.

Lubbock, Constance A., ed. *The Herschel Chronicle: The Life-Story of William Herschel and His Sister Caroline Herschel.* Cambridge: Cambridge University Press, 1933.

Moore, John T. *A History of Chemistry.* New York: McGraw-Hill Book Company, 1939.

Moyer, Albert E. *Joseph Henry: The Rise of an American Scientist.* Washington, D.C.: Smithsonian Institution, 1997.

Newton, David E. *The Chemical Elements.* New York: Franklin Watts, 1994.

Nye, Mary Jo. *Before Big Science: The Pursuit of Modern Chemistry and Physics, 1800-1940.* Cambridge, MA: Harvard University Press, 1999.

Pais, Abraham. *Inward Bound.* New York: Oxford University Press, 1986.

Patterson, E. C. *John Dalton and the Atomic Theory.* Garden City, NY: Doubleday and Company, 1970.

Purrington, Robert D. *Physics in the Nineteenth Century.* New Brunswick, NJ: Rutgers University Press, 1997.

Riedman, Sarah R. *Trailblazer of American Science.* New York: Rand McNally, 1961.

Romber, Alfred. *The Discovery of Radioactivity and Transmutation.* New York: Dover Press, 1964.

Schopf, James. *Cradle of Life.* Princeton, NJ: Princeton University Press, 1999.

Seeger, Raymond John. *Josiah Willard Gibbs: American Physicist Par Excellance.* New York: Pergamon Press, 1974.

Smith Crosbie. *The Science of Energy: The Construction of Energy Physics in the Nineteenth Century.* Chicago: University of Chicago Press, 1999.

Spangenburg, Ray, and Diane K. Moser. *The History of Science in the Nineteenth Century.* New York: Facts on File, Inc., 1994.

Thomas, John Meurig. *Michael Faraday and the Royal Institution: The Genius of Man and Place.* Bristol, UK: Adam Hilger, 1991.

Wheeler, Lynde Phelps. *Josiah Willard Gibbs: The History of a Great Mind.* Woodbridge, CT: Ox Bow Press, 1999.

Whittaker, Edmund. *A History of the Theories of Aether and Electricity.* New York: Harper & Brothers, 1951–1953.

Williams, L. P. *Michael Faraday: A Biography.* New York: Basic Books, 1965.

Web sites

"Faraday, Michael (1791–1867)." *Eric Weisstein's Treasure Trove of Scientific Biography.* [Online] http://www.treasure-troves.com/bios/Faraday.html (accessed on February 26, 2001).

Joseph Louis Gay-Lussac. [Online] http://www.woodrow.org/teachers/chemistry/institutes/1992/Gay-Lussac.html (accessed on February 26, 2001).

Symons, Lenore. *Michael Faraday (1791–1867)* [Online] http://www.iee.org.uk/publish/faraday/faraday1.html (accessed on February 26, 2001).

O'Connor J. J., and E. F. Robertson. "Caroline Lucretia Herschel." School of Mathematics and Statistics, University of St. Andrews. [Online] http://www-history.mcs.st-andrews.ac.uk/history/Mathematicians/Herschel_Caroline.html (accessed on February 22, 2001).

Rotherberg, Marc. "Joseph Henry." *The Joseph Henry Papers Project.* [Online] http://www.si.edu/organiza/offices/archive/ihd/jhp/joseph01.htm (accessed February 26, 2001).

chapter five # Technology and Invention

Chronology **298**
Overview **299**
Essays **303**
Biographies **352**
Brief Biographies **366**
Research and Activity Ideas **372**
For More Information **373**

CHRONOLOGY

1801 French inventor Joseph-Marie Jacquard invents a mechanical loom that operates according to instructions stored on punch cards.

1803 The first steam engine is invented by English engineer Richard Trevithick.

1804 French inventor Nicolas Appert develops the process of canning to preserve food, and opens the first canning factory.

1807 American inventor Robert Fulton builds the *Clermont,* the first steamboat to operate successfully on a commercial basis.

1812 The first automated printing press is designed by German inventor Friedrich König.

1826 French inventor Joseph Niepce produces the first permanent photograph.

1831 American inventor Cyrus Hall McCormick invents his mechanical reaping machine.

1839 American inventor Charles Goodyear accidentally spills a mixture of rubber and sulfur on his wife's stove, discovering the process known as vulcanization.

1844 Having patented the telegraph in 1837, American inventor Samuel F. B. Morse successfully transmits the first Morse code message over a telegraph circuit between Baltimore and Washington: "What hath God wrought?"

1856 English engineer and inventor Henry Bessemer announces his invention of a powerful and inexpensive new method for producing steel.

1862 French inventor Jean-Joseph-Étienne Lenoir builds a vehicle powered by an internal combustion engine.

1876 American inventor Alexander Graham Bell invents the telephone.

1876 Karl Paul Gottfried von Linde builds the first practical refrigerator.

1877 American inventor Thomas Alva Edison builds the first phonograph.

1879 English inventor Joseph Wilson Swan and Edison simultaneously produce the first practical incandescent light bulb.

1885 German engineer Karl Friedrich Benz builds what may be considered the first true automobile, a vehicle which is powered by a gasoline-burning internal combustion engine.

1890 American inventor Herman Hollerith builds the first mechanical calculator using principles on which modern computers are based.

1895 Auguste and Louis Lumière invent the cinematograph, a portable hand-cranked camera that can shoot, print, and project motion pictures. Historians date this year as the birth of the motion picture.

OVERVIEW

Chapter Five
TECHNOLOGY AND INVENTION

The nineteenth century was truly a turning point in human history. Many of the machines, processes, devices, and systems we think of today as being "modern" grew out of nineteenth century inventions and discoveries. Rapid transportation and communication, crowded urban centers, and the market economy are all products of nineteenth century ideas. Even the symbol of the twenty-first century—the computer—developed out of two nineteenth century inventions, the punch card system of Joseph Jacquard (1752–1834) and the analytical engine of Charles Babbage (1791–1871).

Marriage of science and technology

One of the basic characteristics of the nineteenth century was the interaction between science and technology. In this case, science refers to the study of nature for the sole purpose of advancing knowledge. By contrast, technology refers to the invention of specific devices to make some kind of work easier and more efficient. Throughout the nineteenth century, as throughout all of history, science and technology depended on each other.

In some cases, technology took the lead. Someone would invent a method or a device without really knowing how it worked. For example, the invention of canning preceded an understanding of bacteriology. In other words, inventors discovered how to prevent food from spoiling without knowing why it spoiled.

At other times, science took the lead. For example, scientists understood electromagnetism before the telephone was invented. They could describe the way electricity and magnetism were related to each other without knowing how this knowledge could be put to practical use. What scientists and inventors shared was the desire to overcome everyday problems in order to make human life better.

It's a small world after all

The greatest advances that took place during the 1800s affected two areas, transportation and communication. As populations continued to increase, nineteenth-century inventions helped make the world shrink. The first

Technology and Invention

OVERVIEW

locomotive powered by high-pressure steam was built in England in 1803 by Richard Trevithick (1771–1833). A decade later, in 1825, the first commercial railroad system was built to carry steam engines. By 1869, railroad tracks crossed the entire United States from east to west. The time required for a trip across the country was reduced from a few months to a few days.

The invention of the steamboat by American Robert Fulton (1765–1815) in 1807 changed both transportation and politics. Early in the century, steamboats made it possible for people and goods to travel across continents. As steamboats became larger and faster, they were able to cross longer distances. As a result, the countries that were most powerful at the beginning of the century quickly came to dominate the world economy. The impact was most notable for Great Britain, which increased its hold on countries in Africa and Asia. Many times, countries were taken by force as steamboats were able to carry not only people and goods, but also cannons and other weapons.

In some ways, the invention that most propelled the transportation age was the internal combustion engine. The internal combustion engine was invented by Jean-Joseph-Étienne Lenoir (1822–1900) in 1859 and improved by Nikolaus August Otto (1832–1891) in 1877. It provided a compact, light, and efficient system of power for automobiles. Lenoir installed his engine in the first automobile in 1862. By the end of the century, manufacturers in both Europe and the United States were producing thousands of automobiles for the general public.

The communication industry saw comparable leaps during the nineteenth century. The telegraph, patented in 1837 by Samuel F. B. Morse

An illustration of many of the most influential inventions of the nineteenth century, the steam engine, steam boat, steam-operated printing press, and the telegraph. (Courtesy of the Library of Congress.)

Technology and Invention

OVERVIEW

(1791–1872), made it possible to send messages across wires sunk under oceans and strung across continents. Another invention, the telephone, was only a curiosity when Alexander Graham Bell (1847–1922) exhibited it at the 1876 Centennial Exposition. Within two decades, it had become a common sight in homes and businesses. Fast and inexpensive written communication was made possible by three advances in publishing: automated paper production, the steam-powered printing press, and the Linotype machine.

Entrepreneurship is born

Nineteenth-century innovators produced not only the many conveniences enjoyed by millions today, they also brought about a revolution in the way society is organized and the way it operates. For example, the 1800s saw the birth of entrepreneurship, a system in which an individual anticipates trends in business and invests time, energy, and money in the hope of accumulating a large fortune. That system is the very foundation of capitalism, which encourages competition and a free economic market.

Early entrepreneurs include inventors like Nicolas Appert (1749–1841). In addition to inventing new ways to preserving food, Appert also developed the first factories for canning food. He invested much his own money to ensure quality products and, in the process, created the modern food industry. Some inventors, such as Isaac Singer (1811–1875) and Alexander Graham Bell, amassed fortunes from their inventions. Others, like Charles Goodyear (1800–1860), died in debt. But they all shared the drive, determination, and vision that are characteristic of entrepreneurs today.

The process of invention also opened up whole new ways of doing business. The cost of many innovations, such as railroads, subways, and telegraph lines, was enormous. No single person could afford to build them. To obtain the necessary funds, a stock market was created. A stock market allowed many people to pool their money to support new projects. In return, those people expected to see a return on their investments. This system of pooled resources, pooled risk, and pooled profit now dominates the economies of most countries in the world.

New inventions also made possible a much wider system of trading. Products were no longer sold in one town, one county, or even one country. For example, Great Britain, the world's first industrialized nation, exported its manufactured goods to China, Australia, South America, and Canada. It also imported agricultural products from these and other countries to feed Britain's industrial workers The international system of business we know today grew out of this nineteenth century beginning.

Technology and Invention

OVERVIEW

Making life easier, safer, and more entertaining

Many nineteenth-century inventions contributed to urbanization. As more inventions and automatic processes were created, more factories were built. These factories were usually clustered in industrial city centers. More factories meant more job opportunities, and people began to move off farms, out of the country, and into cities and towns. As populations shifted, new technologies were invented to help these new city workers. For example, refrigeration and railroads kept families supplied with fresh, nutritious food. Subways, bridges, and artificial lighting made cities more livable.

Many nineteenth-century inventions solved basic problems of human survival: preserving food to make it safe to eat, making better clothing, and increasing the efficiency of farming. Other inventions achieved a different objective: making life more enjoyable. For example, the first cameras, the first motion pictures, and the first phonograph were all developed during the nineteenth century thanks to the extraordinary vision of Louis Daguerre (1787–1851), the Lumière brothers, Auguste (1862–1954) and Louis Jean (1864–1948), and Thomas Edison (1847–1931).

Changes are met with fear and suspicion

Not all technology, however, was met with a positive reaction. While most people were amazed by the new inventions demonstrated at such elaborate forums as the Great International Exhibition of 1851, some people remained wary. Were they useful tools, or did they pose threats to nature and the divine order of the universe? Many people objected to refrigeration on the grounds that God alone has the power to produce cold. They warned that people who ate previously frozen food were risking their lives.

Robert Fulton's steamboat made it easier for people to travel and for goods to be shipped. (Reproduced by permission of UPI/Corbis-Bettmann.)

Technology and Invention

ESSAYS

Telephones, electric lights, subways, and automobiles were greeted with the same suspicion. Doctors warned their patients against the dangers of driving in a car. Directions for turning on electric lights included the reassuring message, "The use of Electricity for lighting is in no way harmful to health, nor does it affect the soundness of sleep."

Some workers felt that technology threatened their usefulness and their livelihoods. As machines automated the work that was previously performed by skilled laborers, thousands of workers were displaced.

In some cases, a single invention could be both beneficial and harmful. For example, research into the chemistry of explosives led to the invention of dynamite by Alfred Nobel (1833–1896). Dynamite helped to reshape countries by carving out roads and tunnel systems. But it was also used to develop weapons that led to wars marked by massive death and destruction.

Many nineteenth-century men and women of vision helped shape modern society. By attempting to solve everyday problems, they created transportation systems that allow us to travel to far-off places. They allowed us to communicate in the blink of an eye. They developed devices to make work easier. They changed the very nature of how we do business, how we interact, and how we live.

ESSAYS ■

■ THE HORSELESS CARRIAGE: INVENTION OF THE FIRST AUTOMOBILES

Overview
During the nineteenth century, inventors used the steam engine and the internal combustion engine to power the first horseless carriages. By the end of the century, an automotive industry was beginning to develop that would eventually impact every person in every country of the world.

Background
Over the centuries, the way people and goods were transported on land changed very little. People walked, rode horses, or traveled in carts pulled by horses or other animals. By the mid-1400s, inventors such as Leonardo da Vinci (1452–1519) were designing vehicles that were self-propelled. Some inventors built working models that were powered by springs, internal clockworks, or the wind. None of these early experimental vehicles was very practical or reliable.

Words to Know

assembly line: A series of devices, materials, and workers that allow one object after another to be assembled step-by-step.

boiler: A large metal container in which water is boiled to produce steam.

chassis: The supporting frame, or "body," of a car.

coal gas: A type of fuel produced by heating coal in the absence of air.

internal combustion engine: An engine that produces motion when a fuel is burned inside the engine itself.

As the steam engine was developed during the early 1700s, it began to revolutionize transportation. In 1769, French inventor Nicolas-Joseph Cugnot (1725–1804) used steam to power what is considered to be the first working self-propelled vehicle. Designed as an artillery carriage, the vehicle was a three-wheeled cart mounted with a water tank, called a boiler. Kerosene was used to heat the water and pressure from the escaping steam activated the car's mechanisms. Cugnot's carriage could carry four passengers for twenty minutes at a speed of about two miles per hour. This speed is the equivalent of a person walking.

Steam-powered automobiles were used in Great Britain during the first half of the nineteenth century for public transportation. Some of these automobiles were capable of traveling at speeds of twenty miles per hour and could carry up to twenty-two passengers. They were also smelly, noisy, difficult to operate, and sometimes, dangerous. For example, hot sparks from the boiler frequently set fire to crops and wooden buildings and bridges.

By the end of the 1800s and into the 1900s, there were more than one hundred manufacturers of steam-powered automobiles in Europe and the United States. The public, however, never truly embraced the steam automobile. The steam boilers that propelled the vehicles were very large and heavy, and it took quite some time for the water in the tank to heat up. In addition, there was constant fear of explosion. By 1929, steam car production was discontinued.

An alternative to steam was using fuel to power an internal combustion engine. An internal combustion engine operates by burning a fuel inside of the engine itself. It usually runs on gasoline, but it can also operate on other fuels, such as natural gas or kerosene. Internal combustion engines are much smaller and lighter than steam boilers.

The first person to design a non-steam-powered automobile was French inventor Étienne Lenoir (1822-1900). In 1862, he built a primitive automobile powered by an internal combustion engine that operated on illuminating (coal) gas. Coal gas was being used at the time in city and home lighting systems. Lenoir's automobile was powered by a one-cylinder engine and could travel about two miles in six hours.

In 1885, the first automobile of modern design was built by German inventor **Karl Benz** (1844–1929; see biography in this chapter). He mounted a one-cylinder engine on a three-wheeled buggy. Like modern

Technology and Invention

ESSAYS

Karl Benz sitting in his first automobile. (Courtesy of the Library of Congress.)

Technology and Invention

ESSAYS

cars, the engine operated on gasoline. Benz's first vehicles reached a top speed of about eight miles per hour. A driver steered the car using a tiller (a stick-like handle). Later the same year, German inventor Gottlieb Daimler (1834–1900) mounted a gasoline-powered engine on a wooden bicycle to create the first motorcycle.

Benz and Daimler were not the only inventors working on gasoline-powered vehicles. In France, the firm of Panhard-Levassor was the first to mount a gasoline engine in the front of a vehicle in 1891. And in 1898, Louis Renault (1877–1944) was the first to adapt a drive shaft to transfer engine power to the drive wheels.

Great strides were also taking place in the United States. In fact, some historians say that the person who should really get credit for building the first motorcar is the American inventor George Baldwin Selden (1846–1922). In 1879, Selden was given a patent for a gas-powered vehicle design. The date of that patent is about six years earlier than Benz's first working model.

Henry Ford seated in his first automobile. (Reproduced by permission of Corbis-Bettmann.)

306 Science, Technology, and Society

Others claim that the first U.S. gas-powered automobile was produced in 1893 by James F. (1869–1967) and Charles Edgar Duryea (1861–1938) of Springfield, Massachusetts. The two brothers were bicycle mechanics who went on to win the very first automobile race held in the United States in 1895. In 1896, they sold the first commercially produced car in the United States.

During that same year, other automobile innovators, including Henry Ford (1863–1947) out of Dearborn, Michigan, and Ransom Eli Olds (1864–1950), from Lansing, Michigan, introduced their own versions of the first gasoline-powered cars. By 1898, there were 50 automobile-manufacturing companies in the United States; that number rose to over 200 by 1908. In 1899, these manufacturers produced approximately 2,500 motor vehicles.

Impact

By the end of the 1800s, automotive manufacturing was still a fledgling business and very few individuals owned their own cars. Those who did were wealthy because the first automobiles were quite expensive. For example, the 1901 Oldsmobile sold for $650, which was more than the average annual wage earned by a worker in the United States. The average hourly wage in 1900 was fifteen cents per hour; the average yearly salary was less than $500.

Early automobiles were expensive for consumers because they were expensive to manufacture. In Europe, Daimler employed over 1,700 workers to produce less than 1,000 cars per year. In the United States, however, auto makers were beginning to revolutionize automobile production. Ransom E. Olds was the first to introduce a method of mass producing cars in 1901. By 1908, Henry Ford further impacted car manufacturing by introducing the assembly line method of production.

In an assembly line, the skeleton of a car rests on a moving belt. The belt passes by different workers, each of whom is responsible for a single task. With the assembly line, a manufacturer could produce hundreds of cars in a single day instead of just one or two. Once other American manufacturers began to adopt Ford's methods, the price of automobiles began to decrease. By the 1920s, a middle class worker could afford to buy a Ford Model T, which sold for about $290.

Although the automobile was first developed in Germany and France, the United States quickly dominated the automobile industry. By 1927, Ford had sold fifteen million Model T's. By 1929, U.S. companies were responsible for producing approximately eighty percent of the world's automobiles.

As more people bought automobiles, society began to change. For example, the United States covers a very large land area, much larger than

Technology and Invention

ESSAYS

Technology and Invention

ESSAYS

the average European nation. Prior to the invention of the automobile, it was practically impossible for anyone to visit family and friends who did not live within a few miles. Most people worked very close to where they lived and a vacation away from home was virtually unheard of. Automobiles helped to create a much more mobile society.

People in rural areas especially benefitted from automotive vehicles. Doctors traveled by car, which was a faster way of reaching patients in remote locations. Automobiles helped deliver produce to markets; some farmers even used cars to plow their fields.

The automotive industry eventually became so vast that it transformed nearly every aspect of society. In particular, the construction of millions of cars led to a boom in all types of industry, including steel, aluminum, plastic, and rubber. In the twentieth century, the automobile, which was just a vision at the end of the nineteenth century, became a phenomenon that impacted nearly every person on Earth. Today, six million cars are built in the United States each year.

■ REVOLUTIONS IN THE PUBLISHING INDUSTRY

Overview
The Industrial Revolution that spanned the late eighteenth century and the early nineteenth century caused major changes in all types of industry. In the field of publishing, innovations occurred in three areas: the manufacture of paper, the production of type, and the actual printing process itself. Innovations in these areas made it possible to produce greater quantities of printed material at low cost. The impact was especially felt in the newspaper industry.

Background
Technology that came out of the Industrial Revolution changed every step in the publication of printed material. The most important invention was the steam engine. Machines powered by steam engine could do most of the tasks previously carried out by many workers who labored long hours. All forms of production became faster, less expensive, and more efficient.

Paper making is an example. At one time, paper making was a long and difficult task. Wood, rags, and other materials had to be broken down into pulp. The pulp was then mixed with water. The pulp/water mixture was treated to produce the color and texture of the desired product. It was then passed through a fine screen, leaving behind damp sheets of paper. Finally, the paper had to be dried, cut, and stacked.

Words to Know

cottage industry: A business operated, usually in a private home, by an individual, a single family, or group of related individuals.

Industrial Revolution: Change that came about in western Europe in the late eighteenth century when power-driven machinery was introduced to replace hand labor for the production of many objects and materials.

linotype: A machine in which individual pieces of type are cast when an operator types a letter, number, or symbol into a keyboard.

literacy: The ability to read and write.

pulp (as for paper): A raw material, made from wood, rags, wastepaper, and other materials, for use in the manufacture of paper.

steam engine: An engine which is powered by steam.

stick (printer's): A metal form in which individual pieces of type can be arranged to produce a series of words.

type (as in printing): Metal forms that contain the letters, numbers, and symbols from which printed pages are produced.

The first modern paper-making machine was invented in 1799 by a French inventor named Nicholas Louis Robert (1761–1828). All of the steps that it took to make paper were automated, from mixing the pulp and water to pressing and drying. The result was that Robert's machine could quickly create one long continuous roll of paper.

Robert's work had been financed by two English brothers, Henry (1766–1854) and Sealy (d. 1847) Fourdrinier. They later hired an English engineer, Bryan Donkin (1768–1855), to improve on Robert's machine. Donkin received a patent on his improved model in 1805 and built the first working machine two years later. The Fourdrinier machine, as it later came to be called, was first used in the United States in 1816. Machines of this design are still used in paper-making today.

Technology and Invention

ESSAYS

A second stage of mass publishing to benefit from the Industrial Revolution was the production of type. The term *type* refers to the letters, numbers, and symbols that are used to produce printed pages. At one time, each individual piece of type had to be made by hand. Originally, type was carved out of wooden blocks. Later, molten lead was poured into forms to produce type. After the individual pieces of type were created, typesetters manually manipulated them to form words, sentences, and whole pages.

In 1884, German-born American inventor Ottmar Mergenthaler (1854–1899) invented the first mechanized typesetting machine, the linotype. Mergenthaler's machine consisted of a keyboard similar to a typewriter keyboard that had ninety keys. As the machine's operator pressed a key, a copper cast of the letter or symbol (called a *matrix*) was lifted into place. Once the operator finished a line of type, a quick-cooling metal was poured over the casts. When cooled, they would form a complete line of type, or *slug*.

Mergenthaler's linotype machine drastically reduced the manpower needed to create a page of type. It also speeded up the process so that typesetters could set more than 5,000 pieces of type per hour, as opposed to 1,500 per hour by hand. Mergenthaler's machine was so revolutionary that it is considered the most important contribution to the publishing industry since Johannes Gutenberg (c. 1395–1468) invented movable type in the mid-1400s.

As with paper making and typesetting, the actual process of printing was long and tedious. The lines of type had to be set into the printing press. Each piece of paper to be printed was placed into the machine one sheet at a time. A printer then had to manually raise and lower the *platen*, a flat metal sheet that pressed the paper against inked type.

Newspaper men use linotype typesetting machines. The linotype was the first mechanized typesetting machine, invented by Ottmar Mergenthaler in 1884. (Courtesy of the Library of Congress.)

Technology and Invention

ESSAYS

The steam engine provided a way of speeding up this process. The first automated printing press was designed by German inventor Friedrich König (1774–1833) in 1812. König replaced the flat platen with a cylinder that rotated. As it rotated, it carried sheets of paper that were pressed down on inked type. König' machine was first used by the *London Times* in 1814. It printed 1,100 pages per hour, four times the rate of manually operated presses.

The next improvement in the printing press was developed by American inventor Richard Hoe (1812–1886) in 1846. Hoe built a machine in which the type-bed and the paper rolled simultaneously on cylinders. Before that, a type-bed held type in a flat position. Hoe's invention improved the speed of printing even more. In 1847, Hoe's machine, called a rotary press, was put to work at the *Public Ledger* in Philadelphia, Pennsylvania. By 1850, an improved version of the rotary press could print a continuous roll of paper on both sides, and was capable of producing over 18,000 pages per hour.

Impact

Prior to the 1400s, most books and manuscripts were written by hand. In the 1440s, Gutenberg invented movable type and created new possibilities in the printing industry. Little advancement, however, took place between 1450 and the nineteenth century. Each step in the printing process continued to be carried out by hand and the process was so slow that very few books could be produced. Those that were produced were very expensive and only a few wealthy individuals owned small libraries.

By the mid-nineteenth century, however, advances in the publishing industry made printed material readily available at a very low cost. Daily newspapers were established in nearly every part of the Western world. They could be sold at a few pennies per copy, well within the range of many citizens. Magazines and books were also being mass produced. One of the most popular forms of literature at the end of the 1900s was the "penny novel." The penny novel was a short book written on paper of poor quality, but available for only a few cents.

As printed material became more available to the general public, literacy rates increased. The ability to read was no longer a privilege enjoyed by the wealthy. By the beginning of the twentieth century, mass publishing made mass education a possibility in many countries of the world.

The machines developed by Fourdrinier, Mergenthaler, König, and Hoe also brought about significant changes in the publishing industry itself. Prior to the nineteenth century, many stages in the printing operation were carried out by individuals or very small companies. Paper-making, for example, was usually handled by a single family or a very small

group of people. It was known as a "cottage industry." The new machines invented in the 1800s were expensive and small companies simply could not afford to buy them. As a result, most small companies were not able to survive. They were replaced by a few large companies with the resources to compete in a new fast-paced industry.

Technology and Invention

ESSAYS

■ INSTANT MESSAGING: THE INVENTION OF THE TELEGRAPH

Overview
Since the beginning of time, humans have searched for ways to send messages over long distances. In ancient times, some of these methods included using fire to create smoke signals, drumming, and long-distance runners. During the nineteenth century, the development of the telegraph revolutionized communication and made it possible for people separated by an ocean to send a message in a matter of seconds.

Background
The first successful telegraph was developed in 1790 by French engineer Claude Chappe (1763–1805), who had been experimenting with efficient ways for armies to communicate. Chappe's telegraph system, called the semaphore, used visual signals to send messages. The system consisted of tall vertical posts that supported moveable crossbars. Using pulleys and ropes, a semaphore messenger would arrange the position of the crossbars to represent various letters of the alphabet, one at a time. An individual positioned on another post a few miles away used a telescope to view the telegraphed message.

Chappe coined the word telegraph to describe his new system. It comes from the Greek words *tele*, meaning "far," and *graphein*, which means "to write." By the early 1800s, semaphore posts were installed all over France, and remained popular until the birth of the new electric telegraph.

During the early 1700s, individuals were experimenting with methods of sending messages using electric current. In 1727, one inventor managed to send an electrical impulse one-sixth of a mile along a length of thread. In a 1753 issue of *Scots Magazine,* a writer described using a static electricity telegraph to spell out messages. At the time, however, too little was known about electricity for these attempts to fully develop.

Several nineteenth century scientific discoveries unlocked the mysteries of electricity and were responsible for advancing modern communication. In 1800, Italian physicist Count Alessandro Volta (1745–1827) created the first source of a continuous electric current. In other words, he had created the first battery. Two decades later, Danish physicist Hans Christ-

*OPPOSITE PAGE
Illustration of the printing presses of the* London News, *1839. Steam-powered printing presses significantly speeded up the printing process and made printed materials less expensive.*
(Reproduced by permission of Archive/Hulton Getty Picture Library.)

Technology and Invention

ESSAYS

ian Oersted (1777–1851) discovered that a wire carrying an electric current will generate a magnetic field around itself. A year later, English physicist and chemist **Michael Faraday** (1791–1867; see biography in the Physical Science chapter) demonstrated the reverse process in which a magnetic field can be used to generate an electric current.

The first person to put these scientific discoveries to practical use was American artist and inventor Samuel F. B. Morse (1791–1872). Morse had traveled to Europe in 1829, and was intrigued by the semaphore towers he had seen. In 1832, on his trip home from Europe, Morse struck up a conversation with fellow passengers about the new discoveries in electricity. He began to think about ways in which an electrical current could be used to send messages. Morse sketched out ideas during the rest of the voyage.

Samuel Morse. (Courtesy of the Library of Congress.)

Morse's first electric telegraph was built in 1835 out of crudely fashioned homemade implements. It consisted of a pair of wires that connected a sender and a receiver. The sender introduced a current of electricity along the wire. He periodically broke the current to transfer a coded message. The message was then recorded at the receiving end by a pencil, which was moved by an electromagnetic receiver.

Because he had no scientific background, Morse was frustrated by his early attempts to build a working model of his innovative idea. For technical advice, he turned to his friend, Leonard Gale, who was a science professor. Gale introduced Morse to **Joseph Henry** (1797–1878; see biography in the Physical Science chapter), an American scientist who was experimenting with electromagnetism. In fact, Henry had built an electromagnetic telegraph in 1831. The team worked to improve Morse's telegraph and eventually developed a relay system using a series of electromagnets to open and close circuits along a telegraph wire. This allowed messages to be sent over long distances.

Morse demonstrated his new device in 1837. One person who was impressed by the potential for the new telegraph was a young student named Alfred Vail (1807–1859). In 1838, Morse and Vail became partners. Vail's father owned a large ironworks factory and agreed to help finance the enterprise. Vail was a skillful mechanical engineer, and made many

Technology and Invention

ESSAYS

A telegraph operator taps enemy lines. (Courtesy of the Library of Congress.)

improvements to the design of the telegraph. He also helped to refine the system of codes that eventually became known as Morse code.

In 1843, after several years of persuasion, Morse was granted $30,000 by the U.S. Congress to test his invention. Morse and his team were given only two months to install an experimental telegraph wire between Washington, D.C., and Baltimore, Maryland. On May 24, 1844, Morse sent the first successful telegraph message from the United State Supreme Court building in Washington to his partner, Alfred Vail, in Baltimore: "What hath God wrought!"

Impact

Morse's telegraph was an immediate success, and by 1846, telegraph lines had sprouted up across the United States. By 1851, there were over fifty private telegraph companies. In 1856, these companies merged to form the Western Union Telegraph Company.

A telegraph line running across the United States was completed in 1861. It then became possible to send messages instantaneously from one

Technology and Invention

ESSAYS

coast to the other. Five years later, a submarine cable was laid down on the floor of the Atlantic Ocean, connecting North America with Great Britain.

For nearly four decades, the telegraph dominated modern communications systems. Journalists could report on events almost immediately as they occurred. The telegraph allowed businessmen to exchange information quickly. This led to corporate expansion, as branches popped up in distant locations. Most importantly, individuals from anywhere in the world felt connected because instant messaging was a reality.

Many revisions were made to Morse's invention over the years. Improved insulation methods protected overhead and underground lines, which led to more reliable service. A duplex system was developed in Germany, which made it possible for messages to travel simultaneously along a single line. In 1874, American inventor **Thomas Alva Edison** (1847–1931; see biography in this chapter) invented the quadruplex. This allowed for two messages to be sent in each direction at once.

Although the telegraph was eventually replaced by faster communication systems, its impact on society was tremendous. In the nineteenth century, it allowed people to send messages in the blink of an eye. It also paved the way for future communication developments, including the telephone, radio, and television.

■ ALEXANDER GRAHAM BELL INVENTS THE TELEPHONE

Overview

The invention of the telegraph at the beginning of the nineteenth century changed the way humans send messages over long distances. It was the development of the telephone by **Alexander Graham Bell** (1847–1922; see biography in this chapter) in 1876, however, that completely revolutionized the future of communications.

Background

Prior to the 1800s, researchers had experimented with devices that would transmit voices and sounds. In fact, the word *telephone* was coined to describe these early systems. Telephone comes from Greek words that mean "far sound."

The first working telephone was developed in 1863 by German inventor Johann Philipp Reis (1834–1874). Reis's telephone consisted of two mechanisms connected by a wire. The transmitter consisted of a thin membrane and a metallic strip. When sound passed through the membrane, it vibrated. The vibrations caused an electric current to travel across the connecting wire. The receiver was a sounding box made up of an iron needle surrounded by an electromagnetic coil. The electric current from the transmitter caused the iron needle to vibrate.

Reis considered his invention to be scientific toy that was only good for demonstrating the nature of sound. He saw no practical application for it. A decade later, two American inventors, Elisha Gray (1831–1901) and Alexander Graham Bell, saw a very practical use for the telephone.

During the 1870s, Gray and Bell were separately experimenting with ways to improve the telegraph. They were also investigating ways to transmit sound. In 1874, Gray invented an early form of the telephone that consisted of a receiver capable of reproducing tones. He then experimented with the idea of a telephone that would include both a transmitter and a receiver. He did not, however, actually build his design. Instead, in 1876, he filed a caveat with the U.S. Patent Office. A caveat means that someone intends to patent an invention, but the idea has not yet been perfected. It is filed to stop anyone else from using an idea.

Technology and Invention

ESSAYS

Alexander Graham Bell opening the New York–Chicago telephone line. (Reproduced by permission of the Bettmann Archive.)

The First Switchboard Operators

As more people began to use telephones, the most practical way to connect multiple users was through a central switchboard. Within a small geographic region, each individual telephone line had a separate socket, called a jack, on the switchboard. If someone wanted to make a call, the switchboard operator had to make the connection. He or she used short, flexible circuits, called cords, with a plug attached to each end. By inserting one end of the cord into the caller's jack and the other end of the cord into the receiver's jack, the two callers were connected.

The very first switchboard operators were young men, known as "telephone lads." They proved to be unruly workers, however, and soon companies began to hire young women. The first woman operator was Emma M. Nutt, who was hired in 1879 to work in Boston. Women worked long twelve-hour shifts and handled hundreds of calls per hour. Calls came in so swiftly that operators worked the board with both hands. The women workers had to follow a strict dress code and their work was watched closely by supervisors. Despite the hard working conditions, women operators proved to be so efficient that they soon took over the telephone operator job market.

This was an amazing feat considering the time period. At the end of the nineteenth century, job opportunities for women were limited. Most women still did not work outside the home. Approximately sixty percent of those who did have jobs worked as domestic help (as cooks or maids). Some women were teachers, while others worked in factories or workshops. By 1900, less than ten percent worked in offices. Jobs that required the most skill and which paid the most were reserved for men.

In one of the strangest stories in scientific history, Gray discovered that two hours earlier, Bell had already filed his own patent. Bell had been working on a telephone system similar to the one envisioned by Gray. He believed that if two membrane receivers were connected electrically, a sound wave that caused one membrane to vibrate would, in turn, cause the other membrane to vibrate. Bell and his assistant, Thomas Augustus Watson (1854–1934), built two models of their invention in 1875. Although no actual words were transmitted, they did manage to send faint, speech-like sounds.

On March 10, 1876, three days after he was granted his patent, Bell sent the first message using a telephone. Speaking to his assistant in the next room, the message was: "Mr. Watson, come here, I want you." Soon after, Bell began to demonstrate his telephone to the public. One of the first demonstrations took place at the Philadelphia Centennial Exposition in June 1876.

Bell and Watson continued to make improvements to their invention. On October 9, 1876, they made their first long-distance telephone call between Boston and Cambridgeport, Massachusetts (a distance of two miles). In 1877, Bell, along with Watson and partners Gardiner Hubbard and Thomas Sanders, formed the Bell Telephone Company. In May of that same year, the first telephones for commercial use were installed in offices of customers of the E. T. Holmes burglar alarm company.

Impact

Bell attempted to sell his patent to the Western Union Telegraph Company. The company's president, William Orton, refused and instead purchased the designs of Elisha Gray. The company also hired inventor **Thomas Edison** (1847-1931; see biography in this chapter) to develop a new telephone system. Bell sued Western Union for patent violation, meaning they were using a design that only Bell had the right to use. Western Union argued that Bell was in the wrong and that Gray was the actual inventor of the telephone. The courts eventually ruled in Bell's favor. Throughout the 1880s, however, Bell was caught in the middle of over 600 legal battles with people who tried to challenge his patent.

In the meantime, telephone use was spreading rapidly. The first telephones were used primarily for commercial and business purposes, but soon residential customers began demanding service. The first switchboard phone system was installed in New Haven, Connecticut, in 1878. It served twenty-one residences. A few weeks later, the first telephone directory was issued. It was a single sheet containing fifty names.

The first telephones consisted of a single microphone; the user spoke into it and then transferred it to his ear to listen. This made for very confusing conversations. In 1887, Edison designed a telephone with a separate earpiece and mouthpiece. The first telephones also performed rather poorly and were only capable of connecting two lines that were only a few miles apart. Regardless, the potential impact of the telephone was apparent.

As more people used telephones, they clamored for phone service that would cover greater geographic areas. People in remote rural communities

Technology and Invention

ESSAYS

wanted to contact friends, relatives, and businesses in newly developed urban centers. Businessmen wanted to check in with corporate headquarters and with their customers. The first long-distance telephone line went into service in 1881 between Boston and Providence, Rhode Island. The connection in early long-distance calls was somewhat weak and callers were required to do a lot of shouting and careful listening.

By the end of the nineteenth century, many improvements had been made to the telephone and it was quickly becoming part of everyday life. It soon surpassed the telegraph as the quickest way to send messages and eventually it impacted almost every aspect of society. Advances in telephone technology led to the development of the radio, communications satellites, and computer modems in the twentieth century.

■ CAPTURING LIFE ON SCREEN: THE INVENTION OF MOTION PICTURES

Overview
Nineteenth-century inventors needed to know three things in order to create motion pictures: how to make drawings appear to move; how to project images on a screen; and how to capture realistic images on film. During the 1880s, visionaries combined these three elements to produce the first moving pictures. By the end of the century, a cultural phenomena was born.

Background
Moving pictures are possible because of a peculiarity of the eye known as persistence of vision. Persistence of vision means that the eye "remembers" an image for a fraction of a second after the object has actually disappeared. The ancient Egyptians knew about this phenomenon, but it was not fully explained until 1824 by English physician and scholar Peter Mark Roget (1779–1869).

For example, imagine that ten separate still photographs are shown to a person very quickly. Each photograph is just slightly different from the one before. The observer is allowed only 1/100 of a second to view each picture. When that happens, the image of the first picture remains in the eye when the second picture appears. And the image of the second picture remains when the third picture appears. And so on. The brain never has a chance to realize that the ten pictures are separate from each other. The pictures blur together to form a single image that appears to be moving.

You can see this phenomenon at work by playing with a toy called a "flip book." Each page contains a drawing that is slightly different from the one before and after it. When a person flips through the pages of the

Words to Know

celluloid: A hard, transparent film-like material from which motion picture films were once made.

kinetoscope: An early form of a motion picture projector consisting of a wooden box with a peephole in it, through which a person could look to watch a short film loop projected on one end of the box.

persistence of vision: The tendency of the human eye to "remember" an image for a fraction of a second after the object has actually disappeared.

stop-action photography: A process by which a photograph is taken once every second, or some other unit of time, after which the photographs are combined to produce a running film.

thaumatrope: A device consisting of a cardboard disk with images painted on it, in which the images appear to move when the disk is spun.

zoetrope: A device containing a strip of images affixed around the outside edge of a rotating drum; the images appear to move when the drum is rotated.

book, the images appear to move. Toys of this kind were popular as early as the 1830s. The *zoetrope,* for example, was first produced in the 1860s. It contained a strip of images around the outside edge of a rotating drum. As the drum spun, the images blurred and appeared to move.

The second element needed for motion pictures—projection of images—was not new to the nineteenth century at all. Over one thousand years ago, performers in China and India were putting on shadow puppet shows using paper silhouettes. In the sixteenth and seventeenth centuries, images were projected on a screen by shining a light through a piece of glass on which scenes were painted. These devices were called magic lanterns. Inventors designed rotary disks and gears to create special effects.

French inventor Emile Reynaud (1844–1918) combined the magic lantern with a mechanical motion picture device to create a primitive form of a motion picture projector. By attaching a lamp to a zoetrope, he created bright, sharp images of drawings that appeared to move. Beginning in

Technology and Invention

ESSAYS

1892, Reynaud showed his *pantomimes lumineuses* (illuminated pantomimes) to audiences at his Theatre Optique.

The third element needed to create motion pictures was the invention of the camera. The first permanent photographic image was produced in 1826 by French inventor Joseph Nicéphore Niepce (1765–1833). By 1849, a method for attaching photographic prints on glass plates had been invented by the Langenheim brothers of Philadelphia, Pennsylvania.

Louis Lumière. Louis and his brother Auguste were among the early pioneers of the motion picture industry. (Courtesy of the Library of Congress.)

The final step in the evolution of the motion picture was that a camera had to be invented that could take multiple images. The glass plates used by early cameras did not allow for more than one picture to be taken at a time. The multiple-picture camera was made possible by the invention of a flexible material called celluloid. Discovered by an American minister, Hannibal Goodwin, and developed by American inventor George Eastman (1854–1932), celluloid film could be run through a camera in a continuous strip.

In 1889, a team of researchers at **Thomas Edison**'s (see biography of Edison in this chapter) laboratories in West Orange, New Jersey, used this new film to develop a movie camera called a kinetograph. In 1891, they patented the Kinetoscope. The Kinetoscope consisted of a wooden box with a peephole in it. A person could look through the peephole to see a short film projected on one end of the box.

Many historians date the birth of the modern motion picture to December 28, 1895. On that date, brothers Auguste (1862–1954) and Louis (1864–1948) Lumière showed ten short films to a paying audience of thirty-three people in a Paris theater. The brothers had developed a portable hand-cranked camera that could shoot, print, and project motion pictures. It was called a *cinematographe*. They used the camera to film scenes all over the world, including military parades and scenic landscapes.

Less than a year after the Lumière brothers exhibited their motion pictures, two American inventors, Norman Raff and Thomas Armat, bought the rights to Edison's kinetograph. They modified Edison's design so that it could project pictures on a screen. On April 23, 1896, they showed a selection of Edison's kinetograph movies at the Koster and Bial Music Hall in New York City.

Impact

Audiences were immediately thrilled by the opportunity to see moving pictures. In 1894, the first Kinetoscope parlors opened in New York City. Also called peep show parlors or nickelodeons, they soon sprang up across the United States. People flocked to the parlors and paid a nickel to peer through a hole in a box to see a one-minute film. Going to nickelodeons became a popular fad.

The first movies were not what modern audiences are used to. Early viewers were so amazed to see images moving on a screen that they did not expect great drama. During the 1896 preview at the Koster and Bial Music Hall, part of the program featured waves crashing against the beach. People sitting in the first rows were so surprised that they ducked.

Another of the features included the first on–screen kiss between Broadway actors May Irwin and John Rice. Some members of the audience were outraged at the public display of affection. This response marked the beginning of a long controversy over what is appropriate to depict in a motion picture.

The first motion picture projection camera, invented by the Lumière brothers in 1895. (Reproduced by permission of Archive/Hulton Getty Picture Library.)

Technology and Invention

ESSAYS

The first film studios were also a far cry from the modern day elaborate enterprises. The very first studio was located in Edison's laboratories in New Jersey. It was a box-like room mounted on a turntable so that it could rotate. A skylight let in the sun.

By 1900, there were dozens of film companies in the United States and Europe. The films that were produced were very short, usually about fourteen minutes, and actors were paid as little as five dollars per day.

At the same time, inventors were giving way to film pioneers who began experimenting with film technique. French film director George Méliès is considered the father of the modern narrative film. Trained as a magician, he was the first to use special effects, such as double exposure and fading. Between 1896 and 1906, Méliè produced more than 500 films.

As audiences became more sophisticated they demanded longer films with more elaborate story lines. Motion picture makers responded and by the beginning of the twentieth century, motion picture making had turned into a true industry. By 1910, there were nearly 10,000 movie theaters in the United States.

■ ELECTRICITY POWERS THE NINETEENTH CENTURY

Overview

It is difficult to imagine a world without electricity. No computers. No stereos. No televisions. No electric lights. Yet, until the last part of the nineteenth century, electricity was only a scientific curiosity that had little practical application. Scientists, including English physicist **Michael Faraday** (1791–1867; see biography in the Physical Science chapter) and Serbian-born American inventor Nikola Tesla (1856–1943) experimented with electricity and, in the process, discovered the means to generate and use electricity to forever change the way humans live.

Background

Humans have been aware of electricity for more than two thousand years. The early Greeks discovered that rubbing a piece of amber with wool could produce a static electric shock. Scientists, however, did not begin to seriously investigate electricity and what it could do until the eighteenth century. For example, Benjamin Franklin (1706–1790), was the first person to make the connection between lightning and the flow of electricity.

In 1800, Italian physicist Count Alessandro Volta (1745–1827) managed to create a source of continuous electric current. Called the Voltaic pile, it was the world's first electric battery. A few decades later, American

Words to Know

alternating current (AC): Electrical current that changes the direction in which it flows many times per second.

direct current (DC): Electrical current that always flows in one direction.

electrical generator: A machine for producing electrical current.

electrical motor: A device in which electrical current is used to produce some type of motion.

fossil fuel: A term used to describe coal, oil, or natural gas.

hydroelectric plant: A plant in which the energy of running water is used to generate electricity.

physicist **Joseph Henry** (1797–1878; see biography in the Physical Science chapter) and Michael Faraday separately discovered that electric current could produce mechanical movement. They designed and built the first electric motors.

By 1834, American inventor Thomas Davenport (1802–1851) had improved on the design of the electric motor and was using his motor to operate drills and wood-turning lathes. Davenport's motors were later used to develop the electric railway, electric trolley, and electric printing press.

In addition to his work on electric motors, Faraday also built the first electric generator in 1831. A generator is the opposite of a motor; instead of using electric current to create a mechanical movement, it converts mechanical movement into electric energy. An example is a windmill. By connecting a generator to a windmill, the mechanical movement of the windmill generates electricity.

In 1832, French engineer Hippolyte Pixii (1808–1835) built the first hand-driven generator. Other engineers and inventors, including German engineer Ernst Werner von Siemens (1816–1892) and Belgian-French inventor Zenobe Gramme (1826–1901), made improvements on Pixii's invention. Both of them eventually built factories to manufacture electrical devices.

Let There Be Light!

Gas lighting relied on a series of pipes to transfer coal gas (natural gas was not widely used until the twentieth century). It was a relatively cheap way to provide lighting, and it was easy to regulate light by adjusting the controls on the gas burners. The first gas company was established in 1813 in London, England. By 1819, London had 300 miles of gas lines that serviced 500,000 burners. In the United States, a central gas system was providing light for the city streets and buildings in Baltimore, Maryland, in 1816. By the late 1800s, nearly 1,000 American companies were making gas from coal, and gaslight was commonly being used in private homes.

Gas meters were developed in the 1890s that allowed homeowners to turn gas on or off as needed by inserting coins. This pay-as-you-go system made it possible for middle-class families to afford gas lighting. Gas lighting became so widespread that it was strong competition for electric lighting systems that were being introduced in the 1870s.

The first electric lamps produced very bright white sparks. They were so bright, they were used only in city streets, factories, and very large stores. Two inventors were responsible for developing the incandescent light bulb: Englishman Joseph Wilson Swan (1828–1914) and American **Thomas Alva Edison** (1847–1931; see biography in this chapter). Early incandescent bulbs were very much like the modern bulbs of today.

Prior to the 1880s, most electrical devices were operated by direct current (DC). This situation began to change thanks to Nikola Tesla. In 1883, Tesla developed an electric motor and an electric generator, both of which used alternating current (AC). When he went to work for American inventor **Thomas Edison** (1847–1931; see biography in this chapter), he tried to persuade Edison that AC power was much more practical and that it would be the wave of the future. Edison disagreed, and Tesla left to work on his own.

In 1885, Tesla sold his inventions to American inventor and businessman George Westinghouse (1846–1914). Westinghouse and Tesla worked together at the Westinghouse Electric Company to promote the develop-

Edison truly believed that electric light would change the world. In the early 1880s, he introduced the idea of building large electric generators to provide cheap, reliable electricity to the public. In 1881, he built his first electric power station on Pearl Street in New York, New York. By December 1882, over two hundred Manhattan customers, including individuals and businesses, were living and working by electric light.

Until the turn of the century, electric lighting in the home remained a rarity. People were curious about electric lamps, but few people actually installed electricity in their homes. By 1900, only about ten thousand homes in the United States were powered with electric lights. At the time, the United States had a population of approximately seventy-six million.

There was no doubt, however, that electric lighting was transforming public life. People felt safer walking down city streets and through public parks at night. Students could attend evening classes and citizens could hold social and political gatherings in public halls at night. Businesses extended their days past sundown, which led to longer work days for employees and longer shopping hours for customers. Factories stayed open longer, and night shift work became common. Theaters, which traditionally had depended on open flame lamps to light up the stage, switched to electric lights and reduced the threat of fire.

ment of electrical power. By 1866, the Westinghouse company had installed experimental electrical systems in Great Barrington, Vermont, and Lawrenceville, Pennsylvania. That same year, the company began producing generators, transformers, and electric motors.

In 1893, Westinghouse was given a contract to build an electric generating plant at Niagara Falls. In that same year, Westinghouse used AC generators to provide electrical power for the Columbian Exposition in Chicago. The exposition gave Westinghouse and Tesla the opportunity to demonstrate the potential impact electricity could have on the lives of everyday people.

Technology and Invention

ESSAYS

AC power ultimately proved to be superior to DC power. AC motors were lighter, more efficient, and less expensive to produce than DC motors. In addition, engineers were able to transmit AC current over longer distances than DC current.

Impact

It is impossible to list the many ways in which electrical power changed human society in the nineteenth century. Perhaps the most immediate effect was that the power of electricity made work easier. Prior to the invention of electric motors, most work was done using animal or human labor. Sometimes, water or wind was harnessed. For a short time, machines powered by steam were available. Most of these methods, however, were not very efficient.

Gas lighting on a street in 1911. Gas street lighting became popular in the early 1800s. By the late 1800s gaslight was commonly being used in private homes. (Courtesy of the Library of Congress.)

Technology and Invention

ESSAYS

For example, people who owned mills relied on water wheels to grind corn or saw wood. While water was a reliable source of power, mills always had to be built along rivers, or some other source of water. With electric power supplied by a generator, a mill could be built anywhere, and an electric motor could be used to turn the wheels that operated the saws and grinding mortars.

Electricity offered an unlimited supply of power that was available virtually anywhere, regardless of weather or water supply. It also provided a source of power that could outperform any physical labor. The work that took the power of many men or teams of horses could be replaced by electrically powered devices. It was also easier to control the power supplied by an electric motor. This was especially important in the manufacture of smaller objects, such as sewing machines.

Electric motors were particularly important because they changed the way pumps are operated. Pumps are used in many important systems. They supply water to boilers that make steam for electric generators to bring power to homes and businesses. Pumps bring drinking water to cities and provide water used to fight fires.

People in the nineteenth century quickly realized how vital electrical power would be for society. They also realized it was necessary to ensure that electrical power was constant, reliable, and relatively inexpensive. Beginning in Japan and Germany, farmers banded together to form cooperatives. These cooperatives built power stations and transmission lines to bring electricity to individual farms. The electricity was then used to operate grain crushers, threshers, milking machines, and many other types of farm machinery.

It took longer for people in cities to benefit from electrical power. Most homes were not wired for electricity by the end of the century and inexpensive electric rates did not materialize until the late 1920s. Electrical power did begin to impact businesses. For instance, two inventions common to modern stores, the elevator and escalator, were both developed in the late 1800s.

The development of electrical power system quickly spread into the twentieth century and eventually impacted almost every person in every developed nation. Electrical power systems, however, also began to take a toll on the environment. For example, most electrical generating plants operate either with running water or by burning fossil fuels, such as coal, oil, and natural gas. Excavating for fossil fuels often badly damages the land. And when those fossil fuels are burned, huge amounts of air pollutants are formed.

Technology and Invention

ESSAYS

An early incandescent light bulb. Two inventors were responsible for developing the incandescent light bulb: Englishman Joseph Wilson Swan and American Thomas Alva Edison. (Reproduced by permission of UPI/Corbis-Bettmann.)

THE STEAM-POWERED LOCOMOTIVE TRANSFORMS TRANSPORTATION

Overview

The steam-powered locomotive changed the way people and goods were transported, and was an instrumental force behind the Industrial Revolution. As locomotives were perfected during the nineteenth century, it became apparent that this new mode of transportation would eventually affect the way people all over the world live and do business.

Background

The Industrial Revolution, which spanned from about 1780 to 1880, was a period in history when people changed the way they produced and sold goods. Prior to the Industrial Revolution, most work was done by hand. During the 1700s and 1800s, inventors created machines that replaced human power and animal strength. As a result, society as a whole was transformed.

One of the most important inventions that made the Industrial Revolution possible was the steam engine. A steam engine is a device in which boiling water expands into pressurized steam to perform work. In a steam locomotive, for example, water boiled in the engine produces steam that drives the locomotive's wheels. Although no one person invented the steam engine, the individual most responsible for developing the modern steam engine was Scottish engineer James Watt (1736–1819).

By the late 1700s, many inventors realized the potential for steam engines in driving railroad cars. They were also experimenting with types of railway tracks. Primitive railways had existed since the time of the ancient Greeks. Such systems depended on human labor or animal power to pull cars along tracks that had been carved in stone. In the sixteenth century, raised tracks were used to guide carts in mining and quarrying industries. These early tracks were made of wood.

The first successful steam-powered locomotive to run on rails was invented by English inventor Richard Trevithick (1771–1833) in 1803. He installed a steam engine on a flat railway car that could pull other cars. Trevithick had designed his car to haul heavy loads, such as iron, but his simple locomotive was not able to pull very much weight.

Trevithick, however, had proven that a steam-powered locomotive was possible, and inventors quickly began to experiment and make improvements to his design. By 1829, English inventor George Stephenson (1781–1848) had created the first practical locomotive. Called the *Rocket*, it could pull eight coal cars filled with ore at a speed of sixteen miles per hour. This was an amazing speed at the time.

Technology and Invention

ESSAYS

Words to Know

Industrial Revolution: A change that came about in Western Europe in the middle to late eighteenth century when power-driven machinery was introduced to replace hand labor for the production of many objects and materials.

locomotive: A railroad engine that moves under its own power and is used to pull railroad cars.

middle class: That group of people in society who are less wealthy than the upper class, but wealthier than the lower class.

steam engine: An engine which is powered by steam.

ton-mile: A measure equal to a ton of goods carried a distance of one mile.

As locomotives improved, railway systems began to expand. Early nineteenth-century systems still relied on horses to pull cars carrying heavy loads uphill. By 1825, the first modern railroad system to use steam locomotives began operation in England. It ran from the coal fields at Darlington in the north of England to the town of Stockton on the east coast, where coal was loaded onto ships. In 1829, the first railway system, the Delaware and Hudson, was built in the United States. In 1830, the first railway in the world to rely solely on steam locomotion was opened in England. It ran thirty-one miles between Liverpool and Manchester.

Impact

Railway systems had a major impact on nineteenth-century business and industry. For example, railways made it possible to move goods at speeds that were previously unheard of. Trains could travel in virtually all kinds of weather and could reach areas that, in some cases, were quite remote. And, as locomotives became more powerful they could haul many loads of heavy raw materials, such as iron, coal, or stone. With such improved transportation, products could be made faster and sent to market faster. In addition, businesses could sell to markets that were farther away. This

Technology and Invention

ESSAYS

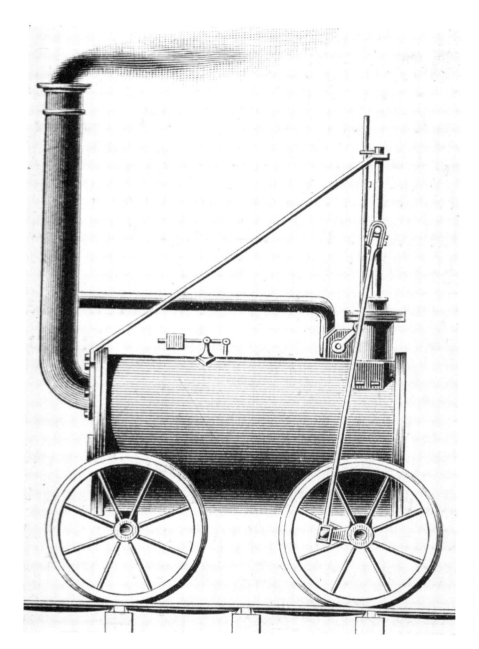

One of Richard Trevithick's steam-powered locomotives, 1804. (Reproduced by permission of Archive Photos, Inc.)

increased the level of economic activity in most countries and greatly expanded trade.

As businesses grew, owners often invested some of their profits in improving railway systems. Soon, a powerful business cycle developed:

Technology and Invention

ESSAYS

A model reconstruction of Robert Stephenson's Rocket, one of the first practical locomotives. (Reproduced by permission of Archive/Hulton Getty Picture Library.)

better railroads led to the improvement of business; and improvements in business increased the need for better railroads.

The railway business gradually became an industry in its own right. Factories were built to manufacture locomotives, railroad cars, track, signals, and switches. Civil engineers were commissioned to plan routes and build bridges and tunnels. Builders were hired to lay miles and miles of new track. Railroads had become big business.

A single railroad cost approximately two million dollars to build in the 1800s. This was an enormous sum of money that no single businessman could afford. As a result, small companies pooled their resources to form corporations. These corporations became very powerful and often controlled local governments, businesses, and properties. Men who led these corporations, including American industrialist Cornelius Vanderbilt (1794–1877), became some of the richest people in the world.

George Pullman Brings Comfort to Passenger Travel

By 1820, some trains in Great Britain were carrying passengers as well as freight. These early passenger cars resembled stagecoaches. They were not enclosed and people crowded together on wooden benches. There was constant danger from smoke and ash spewing from the engine and from car heaters. In addition, the poor design of early cars coupled with uneven railroad tracks made for a very bumpy ride.

Passenger travel became more comfortable thanks to several innovations developed by American inventor George Pullman (1831–1897). In 1865, Pullman introduced the modern sleeping car. He improved on existing designs and added interior decoration and rubber cushions with springs. In 1868, Pullman developed the dining car, which was followed by the first chair car in 1875. Pullman's company also contributed to passenger safety by introducing such features as the vestibule, which allowed passengers to move safely between railroad cars.

Although railways were originally intended to transport goods, by the mid-1800s, trains were carrying passengers. By making travel easier, railway systems contributed to the growth and development of nations. For example, at the beginning of the 1800s traveling from the east coast to the west coast in the United States was very difficult. People traditionally traveled by open wagon or stagecoach. To open up the West, the federal government started to fund the building of railroads in the 1850s. In 1869, the East and the West were connected when the Union Pacific Railroad and the Central Pacific Railroad were joined in Promontory Point, Utah.

By the end of the century, over 200,000 miles of track covered the United States alone. Worldwide, railroads promoted growth in a variety of industries, including coal, steel, and commercial farming. They made it possible for smaller towns to form and large urban centers to develop. Railroads changed the way people traveled and ultimately, triggered changes that dramatically altered Western civilization.

Technology and Invention

ESSAYS

THE MECHANIZATION OF TEXTILE WEAVING

Overview

Concerned about the poor working conditions of weavers in his hometown of Lyon, France, Joseph-Marie Jacquard (1752–1824) invented the first successful automated weaving loom in 1801. Jacquard's loom transformed the textile industry, making it possible for machines to do the work formerly required of many men and women. In addition, because his loom relied on a system of punched cards that stored data, Jacquard ultimately influenced inventors who were developing the first automatic computing machines.

Background

In the 1700s, the weaving of cloth was a very laborious task. A worker sat before a large frame that supported many vertical threads. He or she then passed a piece of wood containing a spool of thread (the shuttle) back and forth through the threads. The desired pattern was produced by lifting certain vertical threads, but not others.

By the late eighteenth century, the demand for more intricately patterned cloth was increasing. This meant more work for weavers everywhere. The demand was particularly felt in the city of Lyon, France, which was a center of silk weaving. Entire families, including children, were involved in the task of weaving. They commonly worked in crowded rooms from dawn to dusk every day of the week. The work was so difficult that children grew old very quickly and adults often died early from overwork.

Jacquard witnessed how silk weavers were living in Lyon, and was determined to invent a machine that would make their work easier. He was not, however, the first person to try to mechanize the process of weaving. For example, Basile Bouchon had invented an automatic system for weaving in 1725. The system used a roll of paper tape that passed over a rotating cylinder. Holes punched in the tape controlled the movement of the weaving shuttle. Each time a hole appeared in the tape, the shuttle was allowed to pass through the vertical threads. When there was no hole, the shuttle was prevented from moving.

Improvements in Bouchon's design were made by French inventor Jacques de Vaucanson (1709–1782). In 1775, Vaucanson built a mechanized loom with a complicated system of needles and hooks controlled by programmed cards punched with holes. The system worked reasonably well, but it was too complicated for practical use.

Jacquard based his invention on the designs of Bouchon, Vaucanson, and others. One of the important improvements he made was to replace

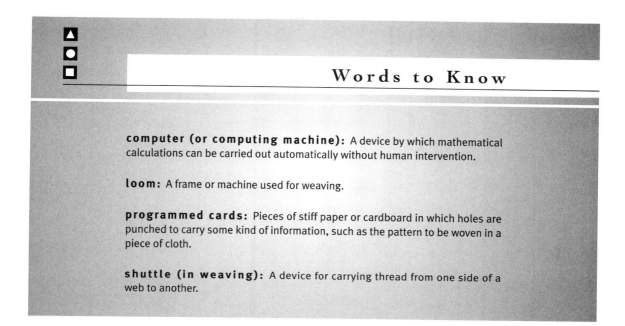

Bouchon's paper tape with very stiff cardboard cards that rotated on a prism. He also automated the movement of the shuttle using a system of hooks and wires.

In the Jacquard loom, cards were punched with holes to control the weaving pattern. In order to create a specific design, a punched card was moved into place in the loom. The weaving needles passed through the holes in the card and certain threads were lifted to make a specific section of the pattern. When there were no holes, the needles were pushed back from the card and no thread was lifted. Using several Jacquard attachments, a weaver could create very intricate patterns over large areas.

Jacquard displayed his loom at the Paris Industrial Exposition in 1801, and was awarded a bronze medal for his work. The French ruler Napoleon I (1769–1821; emperor of France 1804–1815) was so impressed by the loom that he gave Jacquard a pension that would make him financially secure.

Impact

Jacquard's invention revolutionized the weaving of cloth. Only a decade after the Paris Exposition, more than 11,000 of his looms were in use, primarily in France and England. They were adapted for the manufacture not only of silk cloth, but also of cotton, wool, and other fabrics. Although some modifications have been made to it, the knitting and weaving industries throughout the world continue to rely on the Jacquard loom today.

Technology and Invention

ESSAYS

A Jaquard loom. Jacquard's invention was the first to successfully mechanize the weaving process. (Reproduced by permission of UPI/Corbis-Bettmann.)

Jacquard's invention also impacted a totally different field. His idea of using punched cards to program a machine led to the development of the first computers. In 1834, the year Jacquard died, English mathematicians **Charles Babbage** (1791–1871; see biography in Mathematics chapter) and **Augusta Ada Byron**, Lady Lovelace (1815–1852; see biography in this chapter) were using punched cards to create a machine that would automatically compute complicated mathematical calculations. They experimented for years, but neither of them was ever able to produce a working model of a computing machine based on punch cards.

In 1890, American inventor Herman Hollerith (1860–1929) created the first machine that successfully relied on punched cards to read and

Jacquard's Loom Causes Riots in Lyon

In 1801, Joseph-Marie Jacquard (1752–1824) was awarded a bronze medal by the French government for creating an automated loom that would benefit the workers of France. When he returned to his home of Lyon to build his looms, he was not greeted with the same enthusiasm. Weavers were particularly upset because they feared they were going to lose their jobs and be replaced by machines. To protest, workers began to riot and, on more than one occasion, Jacquard was attacked. To calm the rioters, Lyon government officials destroyed one of Jacquard's new looms in the city square.

Protests of this kind were not uncommon throughout the eighteenth and nineteenth centuries, as improved weaving machines threatened to displace skilled weavers. Weaving machines were the first mechanical devices to replace a large number of laborers. To show their displeasure, workers were known to throw their heavy work shoes, called *sabots*, into the weaving machines to destroy them. The word *sabotage*, which means "to deliberately destroy property or stop normal operations," may have been coined to describe such actions.

Workers in Lyon gradually accepted the idea of mechanized looms, especially since the city began to experience a booming economy. Before his death, Jacquard was awarded both a gold medal and the cross of the Legion of Honor for his invention. Six years after he died, the city erected a monument in his honor in the town square. The monument stands on the exact spot where his loom had been destroyed forty years earlier.

sort data. He used his machine to process data for the 1890 U.S. census. By the 1880s, the task of collecting, preparing, and tabulating information about the U.S. population was enormous. Hollerith's machine was used to complete the entire 1890 population count in six months, which was one-third the time it took in 1880.

Hollerith expanded on uses for his system of punched cards and opened his own business called the Tabulating Machine Company. In 1924, it became International Business Machines, more commonly known as IBM.

Technology and Invention

ESSAYS

Jacquard developed his system of punched cards to ease the plight of weavers. In the process, he created a method for storing information that could be used over and over to carry out operations. Punched cards were used to develop the first automated computer in 1944. Computer punch cards were used to store, process, and tabulate data well into the 1970s. How surprised Jacquard would be to know that his innovative idea ultimately helped usher in the modern computer age.

■ PROGRESS IN FOOD PRESERVATION

Overview

Humans have been preserving food for thousands of years. In the 1800s, methods that were developed by the earliest civilizations, such as drying, freezing, fermenting, and salting, were still being used. Throughout the century, many improvements were made in these various food preservation processes.

Background

One of the first preservation methods discovered by early humans was drying. Drying fish, meat, fruits, or vegetables removes water from food. Without water, harmful microorganisms that make food unsafe to eat, cannot survive. Early cultures dried food in the sun and, later, used fire. Using fire to dry foods led to smoking, which is another method of preservation. Foods that were dried could be safely stored for many weeks or months. For example, both the Egyptians and the Romans stored wheat and barley in large silos. Hot dry air trapped in the silos often preserved the grain for years.

In cold northern regions, people kept food fresh by cooling or freezing. They were not aware of it, but cold temperatures halted the growth of harmful microorganisms. In southern areas, keeping food cool was practically impossible. As an alternative, a preferred method of food preservation was fermentation. When a food ferments, it produces acids that prevent harmful bacteria from growing. Rice, grapes, and barley were fermented into alcoholic beverages. People also pickled food, which means it was preserved in an acid-like substance, such as vinegar. Salt was another substance used to preserve food. Salt kills bacteria and also tends to reduce the amount of moisture in foods.

All of these age-old preserving techniques were used for years, but by the nineteenth century, inventors began looking for more efficient ways to keep foods fresh. For example, two French inventors by the names of Masson and Chollet invented a system for drying vegetables in the early 1800s. They cut vegetables into very thin slices and then dried them with

Words to Know

canning: A method for preserving food by cooking it and sealing it in airtight containers.

evaporated milk: Milk that has been heated so that much of the water it contains is drawn off.

fermentation: Any process by which food is changed when acted upon by yeast, bacteria, or enzymes.

food-borne disease: A disease caused by germs present in food that has spoiled.

food preservation: The process of protecting food from decaying by means such as drying, freezing, or salting.

germ theory: The scientific concept that infectious diseases are caused by microorganisms, or germs.

silo: An airtight building used for the storage of grains.

toxic: Poisonous.

vitamin-deficiency disease: A disease that develops when a person does not receive an adequate supply of a vitamin.

working fluid: The liquid or gas used in a refrigeration system to carry heat away from the system.

a mechanical warm air system. Writers of the time claimed that the vegetables retained their flavor for many weeks.

The first machine for drying foods was invented in 1886 by American inventor A. E. Spawn. Named the "Climax Fruit Evaporator," it consisted of slowly revolving trays that were heated by currents of hot air.

Some of the most useful inventions in food preservation were developed by French inventor **Nicolas Appert** (c. 1750–1841; see biography in this chapter). Appert invented several practical methods for drying milk,

Technology and Invention

ESSAYS

meat, and vegetables. He is best known, though, for inventing the process of canning. The system he developed involved three steps. In step one, food was placed into glass bottles, which were then lightly corked. In step two, the bottles were placed into boiling water. In step three, the bottles were removed from the water, and the cork stoppers were inserted tightly into the bottle necks.

Appert did not know it, but his heating system was successful for several reasons. First, the heat from the boiling water was sufficient to kill nearly all bacteria in the food. Second, sealing the bottles tightly prevented air from getting into the bottles. Without air, any bacteria that remained in the food would not be able to grow.

Appert described his process for canning food in a pamphlet printed in 1810. At the same time, an English inventor named Peter Durand had been working on a similar process. Durand's method of canning used tin cans rather than bottles. By 1814, factories in England were producing canned foods for shipment to British troops overseas.

The first Frigidaire refrigerator, shipped from Dayton, Ohio. The ability to refrigerate food revolutionized the food industry. (Reproduced by permission of UPI/Corbis-Bettmann.)

A number of devices for refrigerating foods were also invented during the 1800s. Inventors realized that when a liquid changes to a gas (when it evaporates), it absorbs heat. When a gas returns to a liquid state, it releases heat. By containing this heating and cooling process in a single unit, they invented the first ice-makers and cooling machines. Early refrigeration devices used ethyl ether as the working fluid. Later inventors tried air, sulfur dioxide, and other fluids.

In 1844, American physician John Gorrie (1803–1855) created the first ice-making and air-conditioning machine. By the 1870s, the ice-making industry was flourishing and iceboxes were common in many American homes. In 1876, German inventor Karl Paul Gottfried von Linde (1842-1934) invented the first commercially successful refrigerator for use in German breweries.

Impact

At first, the new and improved methods of food preservation were used primarily to supply foods to people in remote places, such as soldiers, sailors, and explorers. Dried fruits and vegetables and tinned meat and fish provided nourishing foods to untold numbers of men and women who would otherwise have gone hungry, developed food-borne diseases, or suffered from vitamin-deficiency diseases.

Advances in food preservation, however, eventually resulted in a large-scale food industry that created thousands of new jobs. To keep up with consumer demands, factories were constructed to process and pack foods. Cold-storage warehouses were built to house perishable foods that were being transported, and refrigerated boxcars and steamships moved food from farms to cities and from one country to another. The food industry became such a big business that large food empires emerged during this time, including those run by American businessmen George Borden (1801–1874) and Philip Armour (1832–1901).

A high demand for various foods changed the way farms operated. At one time, many farmers grew a variety of products. After the nineteenth century, farmers began to specialize in specific items, such as meat, eggs, or dairy products. This allowed them to increase their production. Improved methods of transporting food allowed farmers to sell to customers who lived far away. This opened up new markets and increased profits.

Food preservation techniques ultimately changed the way of life. At one time, the vast majority of people ate food grown on their own farms or at other farms no more than a few miles away. With effective methods of food preservation, people in the nineteenth century began to purchase food instead of producing it. In addition, their diets were no longer limit-

Technology
and Invention

ESSAYS

ed to foods that were in season. Fresh fruits and vegetables could be enjoyed even during the winter. Eventually, a person could, in theory, enjoy almost any kind of food at almost any time of the year in almost any place on Earth.

■ ADVANCES IN THE CHEMISTRY OF EXPLOSIVES

Overview

Warfare took a deadly step forward during the nineteenth century. A number of powerful new explosives were discovered that made it possible for armies to kill more soldiers much more efficiently than ever before. The new explosives, however, also proved beneficial for society. They were used to carve out new roads and tunnels and to mine and drill for natural resources. Researchers also used their knowledge of explosives to create new fertilizers and medicines.

Background

The oldest known explosive in the world is gunpowder, which was probably discovered in China during the tenth century A.D. Gunpowder is a mixture of potassium nitrate, charcoal, and sulfur. When this mixture is ignited, it produces a powerful blast. Gunpowder is also known as black powder or blasting powder.

Gunpowder was first used to fuel fireworks, rockets, and fire bombs. It was not used in warfare until 1304, when Arabs developed the first primitive gun. These early guns were made from hollow bamboo stalks reinforced with iron; gunpowder was used to fire arrows out of the guns. By the mid-fourteenth century, gunpowder was used to fire cannons, guns, and other military devices.

Over the centuries, scientists experimented with the chemical mixtures that were used to create gunpowder. By the mid-nineteenth century, chemists had become very interested in the science of explosives. In 1845, German chemist Christian Schönbein (1799–1868) accidentally discovered the first new explosive while he was carrying out experiments in his wife's kitchen. In one experiment, he combine nitric and sulfuric acids. When he accidentally spilled the mixture, he grabbed his wife's cotton apron to wipe up the spill. He then hung up the apron to dry. The apron dried, but to Schönbein's amazement, it then blew up in a puff of smoke. He had invented the explosive now known as guncotton.

Schönbein saw the commercial possibilities for his discovery, and opened a factory to manufacture his new invention. It proved to be a very

Words to Know

cordite: A type of explosive made by combining nitroglycerine, guncotton, and vaseline.

dynamite: An explosive discovered by Alfred Nobel that consists of nitroglycerine soaked in diatomaceous earth.

explosive: A chemical which, when shocked or ignited, burns very rapidly and releases a large shock wave.

guncotton: Another name for the explosive nitrocellulose (or cellulose nitrate), made by reacting nitric and sulfuric acids with cotton.

gunpowder: An explosive mixture of potassium nitrate (nitre or saltpeter), charcoal, and sulfur; also known as black powder or blasting powder

nitroglycerine: An explosive made by reacting nitric and sulfuric acids with glycerine.

Nobel Prizes: A group of awards in various fields of science provided for in the will of Swedish businessman Alfred Nobel.

smokeless powder: A general name given to any explosive powder that produces relatively small amounts of smoke when ignited, but referring especially to a form of guncotton with this property.

synthetic fertilizer: Any compound developed for use in agriculture that provides nutrients to crops.

unstable product, however, and was dangerous to manufacture. It eventually fell out of favor.

Other explosives that proved to be even more deadly were being developed at the same time. In 1847, Italian chemist Ascanio Sobrero (1812–1888) combined nitric and sulfuric acids with glycerin to form nitroglycerine. Sobrero discovered that the new compound exploded very easily when heated or shaken. Unlike Schönbein, Sobrero was so horrified by his discovery that he never tried to put it to use.

Technology and Invention

ESSAYS

Other scientists were not so cautious. They saw the potential for nitroglycerine in industrial and military applications and set up factories to manufacture the product. Unfortunately, nitroglycerine is a very sensitive material that explodes very easily. Many people were killed during the years when nitroglycerine was first being produced.

One person who was especially interested in the potential of nitroglycerine was Swedish inventor Alfred Nobel (1833–1896). Nobel's family had long been involved in the manufacture of weapons and explosives. When Nobel began to manufacture nitroglycerine, his own brother was killed in an explosion at one of his factories.

In 1866, while working with nitroglycerine, Nobel had an accident that produced another new explosive. Some nitroglycerine leaked into a container filled with diatomaceous earth. Diatomaceous earth is a powdery material made from the decayed bodies of very small organisms that live in the ocean. Nobel was surprised to discover that the combination of nitroglycerine and diatomaceous earth was very stable. It could be heated and shaken without producing an explosion. The combination could, however, be set off with a smaller explosive blast. Nobel called his invention *dynamite*.

Alfred Nobel, inventor of dynamite.
(Courtesy of the Library of Congress.)

Nobel was also working on other problems involving explosives. Nitroglycerine, dynamite, and most other explosives produce huge clouds of black smoke. On the battlefield, smoke from cannons and guns created a thick screen that blinded soldiers. Military leaders began asking for a smokeless alternative. In 1888, Nobel began production of the first smokeless powder, which he called ballistite. A year later, two British chemists, Frederick Abel (1827–1902) and James Dewar (1842–1923) produced another type of smokeless powder.

The Abel-Dewar product was made by combining nitroglycerine, guncotton, and vaseline. The result was a stable liquid that could be made into almost any shape. Most frequently, it was formed into long, rope-like cords. The cords could be cut to any length, depending on the desired size of explosion. Because of this property, the new product was called *cordite*.

Technology and Invention

ESSAYS

Sticks of dynamite are primed for an explosion. Dynamite, an explosive that combines nitroglycerine and diatomaceous earth, is much more stable than straight nitroglycerine, which explodes very easily. Dynamite has to be set off with a smaller explosion. (Reproduced by permission of Archive/Hulton Getty Picture Library.)

Impact

Explosives discovered in the nineteenth century quickly became popular, and were used for both military and peaceful purposes. By the end of the century, Nobel alone had built factories in many countries and was producing 66,500 tons of dynamite a year. By the end of the twentieth century, about five billion tons of explosives were being manufactured in the United States every year.

The first widespread use of dynamite and other new explosives in warfare occurred during World War I (1914–1918). They were used in artillery shells, bombs, land mines, and other types of weapons. By the end of the war, due in part to explosive technology, eight-and-a-half million people lost their lives.

At the same time, explosives were being used for many other purposes. As populations grew and nations expanded, explosives helped to reshape society. Dynamite was used to clear land for roads and buildings;

Technology and Invention

ESSAYS

to tap into underground sources of oil; to build dams; to excavate for coal, iron, and other minerals; and to dig tunnels for subways and highways.

Research into the chemical properties of explosives also produced unexpected benefits. Scientists working on these materials learned a great deal about the element nitrogen, one of the essential components of explosives. This new knowledge eventually led to the development of nitrogen-based synthetic fertilizers. Fertilizers helped nineteenth-century farmers to increase their food production, and eventually transformed the nature of modern agriculture.

Another byproduct of the explosive business was the Nobel Prize. Known as "the merchant of death," Alfred Nobel made a fortune from explosives. In his will, he directed that nine million dollars be set aside to honor excellence in chemistry, physics, medicine, literature, and peace. Today, the Nobel Prize is considered one of the highest honors that can be bestowed on an individual.

■ CHARLES GOODYEAR INVENTS THE PROCESS OF VULCANIZATION

Overview

Europeans had been fascinated with rubber since it was first discovered by explorers to the New World during the 1500s. Although natural rubber is elastic, moldable, strong, and waterproof, it is also a substance that is affected by temperature. Heat causes rubber to become soft and sticky, while the cold makes rubber turn hard and brittle. In 1839, American inventor Charles Goodyear (1800–1860) discovered a process, known as vulcanization, which transformed rubber into a practical substance with countless uses.

Backgrouond

Natural rubber is obtained from the rubber tree, *Hevea brasiliensis*. It oozes from the bark of the tree in a sap-like form known as latex. People all over the world have long used natural rubber. They produce rubber by drying latex over heat. The heat evaporates the water and leaves balls of latex behind.

While natural rubber was a familiar product in many parts of the world, it was unknown to Europeans until the late sixteenth century. Explorer Christopher Columbus (1451–1506) first encountered the product when he visited the island of Hispaniola (now called Haiti). Columbus saw native people playing with light, bouncy balls made from rubber. He brought samples of the material back to Europe.

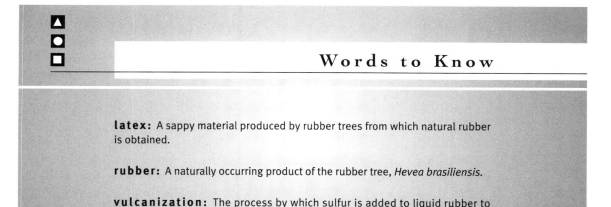

latex: A sappy material produced by rubber trees from which natural rubber is obtained.

rubber: A naturally occurring product of the rubber tree, *Hevea brasiliensis*.

vulcanization: The process by which sulfur is added to liquid rubber to make the rubber stronger and more resilient.

In 1738, French explorer Charles Marie de la Condamine (1701–1774) brought back samples of rubber from South America, and described the various ways it could be used. The new material was at first given the name *caoutchouc*. Caoutchouc is a Mayan word that means "weeping wood." It was also called *gum elastic*. It was eventually called *India rubber* in 1770. "India" because it came from the West Indies, and "rubber" because English chemist Joseph Priestley (1733–1804) found that the material was effective in rubbing out pencil marks.

Scientists and inventors began to study this product from the New World and soon realized that the material had many desirable properties. It was strong, durable, water resistant, and elastic. It seemed that rubber could be used in many different industrial and commercial applications. One of the first of these applications was in the waterproofing of cloth. Natural rubber was first dissolved in a liquid, such as turpentine or naphtha. The solution could then be applied to cloth, making it resistant to water. Rubber was ideal for this purpose because it was flexible and water resistant.

Perhaps the most famous producer of rubber products was a Scottish chemist by the name of Charles Macintosh (1766–1843). Macintosh pressed two pieces of rubberized cloth together, like a sandwich. He then used the material to make waterproof raincoats. Raincoats of this design are still widely known as *macintoshes*.

Other manufacturers were using rubber to make a number of products, and there was an India-rubber craze throughout Europe and the United States. Rubber was used to make items such as coats, hats, shoes, suspenders, and wagon covers. English inventor Thomas Hancock

Technology and Invention

ESSAYS

(1786–1865) manufactured both clothing and footwear from rubberized cloth. Hancock also invented the first successfully commercial rubber manufacturing machines.

For all these developments, rubber proved to be a seriously flawed material. The problem was that rubber was affected by hot and cold temperatures. For example, the raincoats invented by Macintosh worked well in the cool London fog. In warmer climates, they melted into a sticky, smelly mess. At the other extreme, rubber boots worn during a very cold winter turned brittle and crumbled apart.

One of the individuals who attempted to invent a more useful type of rubber was a young American inventor named Charles Goodyear. Goodyear had never been very successful as an inventor or businessman. He was bankrupt so many times that he once referred to debtor prisons as his "hotels." He conducted many of his early experiments in prison kitchens.

Beginning in the mid-1830s, Goodyear tried to change the physical properties of rubber. He took a simple approach. He mixed every material he could think of with rubber to see what effect it would create. Among the substances he used were salt, sugar, witch hazel, castor oil, ink, and cottage cheese. Experiments with magnesia, quicklime, and nitric acid offered only limited success.

Then, in 1839, Goodyear had one of the most famous accidents in the history of science. He was heating raw latex with sulfur on his wife's stove. Some of the mixture accidentally spilled on the stove, and when it cooled, Goodyear made

Charles Goodyear, inventor of the process of vulcanization. (Courtesy of the Library of Congress.)

a fascinating discovery. The mixture remained solid and strong when heated. It also retained its flexibility at low temperatures. In other words, rubber treated with sulfur retained its most desirable properties across a wide range of temperatures. The process discovery by Goodyear was later called *vulcanization*, in honor of Vulcan, the Roman god of fire.

Impact

Word of Goodyear's discovery soon spread through the business world. Manufacturers began making all types of products from vulcanized rubber, including bottle stoppers, frames for photographic plates, cigarette hold-

The Fate of Charles Goodyear

The development of rubber into a useful substance proved very beneficial to society. And, as the demand for rubber products increased, the rubber industry made many manufacturers very rich men. By 1860, more than 60,000 people were working in U.S. rubber plants and the industry was bringing in approximately $8 million per year.

The person who benefitted least from the rubber industry was the man responsible for creating the process of vulcanization, Charles Goodyear. The process that Goodyear invented was such a simple one, that it was very easy for other inventors to copy it. By 1844, Goodyear had patented his invention, meaning he had filed an official notice in the United States that gave him the sole rights to use the process of vulcanization. He did not have enough money, however, to file the patent in Great Britain.

British manufacturers, including Thomas Hancock (1786–1865), quickly discovered how to replicate Goodyear's vulcanization process and filed patents of their own. In the United States, Goodyear constantly battled with inventors who attempted to steal his patent. He hired attorney Daniel Webster (1782–1852) to fight for his rights in court. Goodyear paid Webster $15,000, which was more money than he ever earned from his invention.

To promote his process, Goodyear spent $30,000 to build an elaborate display at the Great Exhibition of 1851. The exhibition was held in London to showcase the most amazing inventions of the day. Every bit of Goodyear's display, including furniture, musical instruments, and balloons, was made from vulcanized rubber. He created another lavish display for the Paris Exhibition of 1855.

None of Goodyear's efforts proved worthwhile. He remained virtually penniless, just as he had been throughout most of his life. At various times, he sold his children's schoolbooks and his wife's jewelry to pay bills. Although he had invented a remarkable process that led to the creation of a worldwide industry, when Goodyear died in 1860, he left behind a debt of more than $200,000.

Technology and Invention

BIOGRAPHIES

ers, rubber dental plates, harnesses, and many items of clothing. Consumption of rubber increased from 38 tons in 1825 to 8,000 tons in 1870.

The biggest boost to the growth of the rubber industry was the invention of the automobile. Early cars used solid rubber tires, although they were soon replaced by air-filled, or "balloon" tires. Rubber was also used to make bicycle tires. The invention of electrical devices, such as the incandescent electric lamp, further increased the demand for vulcanized rubber. The rubber was used to insulate electrical wiring and cables. Another big application for rubber was in making conveyer belts, which were used to move everything from seeds and grain to minerals and ores.

Today, more than two billion pounds of natural rubber are used in the United States alone each year. About three-quarters of that amount is used to make tires. Conveyor belts and electrical equipment account for most of the rest of the consumption.

■ BIOGRAPHIES

■ NICOLAS APPERT (C. 1750–1841)

Nicolas Appert started life as a French chef. He eventually became an inventor, and is credited with developing the modern method of canning food.

Appert was born around 1750 in Chalons-sur-Marne, near Paris, France. His father was an innkeeper, who taught his son about cooking, food preservation, and wine-making. Appert became an apprentice at the Palais Royal Hotel, and was then appointed chef to the Duke and Duchess of Deux-Points.

In 1784, Appert used a small inheritance to open a candy shop and grocery in Paris. He soon became famous for the cakes, candies, and sweets he made. In 1789, when revolution broke out in France, Appert donated money to the revolutionary army. He was arrested and imprisoned briefly for his political beliefs during the chaotic period that followed the revolution known as "The Terror."

During the 1790s, Appert went from chef to inventor. At the time, Napoleon's (1769–1821; emperor of France 1804–1815) armies were advancing throughout Europe in their battles of conquest. By the time supplies were sent to troops in the field, the food was spoiled. In 1795, the French government offered a prize to anyone who could find a way of preserving food. Appert decided to take on the challenge.

After much trial and error, Appert developed a method of canning that successfully preserved food. He found that storing food in containers, then heating the containers in boiling water, kept food from spoiling. By 1802, Appert had purchased a large piece of property in Massy, France, where he established a farm to grow fresh produce. In 1804, Appert opened the first canning factory on the same property. The result was that food was grown and then quickly preserved for freshness.

French naval authorities tested Appert's preserved food on board their ships. After a year at sea, the food remained fresh. Another test was conducted with red wine that, at the time, spoiled quickly when exported to other countries. Appert stored several bottles of preserved wine on board a ship that set sail to the Caribbean. When the bottles returned, experts pronounced the aged wine to be superior to fresh wine.

On January 30, 1810, Appert was awarded the government prize of 12,000 francs for his preservation techniques. A year later, he wrote a book describing his methods, *The Art of Preserving All Kinds of Animal and Vegetable Substances for Several Years*. In 1820, he was given a gold medal for his accomplishments and, in 1822, he was awarded the title "Benefactor of Humanity."

Despite these successes, Appert suffered many setbacks. His canning factories were destroyed in 1814 when Prussian and Austrian armies invaded France. He started over again and built a new canning factory in Paris at the age of sixty-four. By this time, Appert had started to store food in tin cans instead of glass bottles.

When Appert was seventy-eight years old, the French government stopped providing support for his factory. In 1836, he retired to Massy, where he died alone on June 3, 1841. By that time, he was penniless and was buried in a pauper's grave.

Technology and Invention

BIOGRAPHIES

Nicolas Appert. (Reproduced by permission of Brown Brothers.

■ ALEXANDER GRAHAM BELL (1847–1922)

Alexander Graham Bell is known throughout the world as the inventor of the telephone. He also invented a number of other devices, including a primitive tape recorder, a hydrofoil speed boat, and a device to detect metal

Technology and Invention

BIOGRAPHIES

in the human body. In addition, Bell was a speech teacher, who spent most of his life teaching the deaf and developing systems to help the deaf.

Bell was born in Edinburgh, Scotland, on March 3, 1847. His mother was nearly deaf and his father was a speech educator. Bell was taught at home by his mother until he was eleven years old. He then entered the Edinburgh Royal High School. When he was fourteen, he moved to London, England, for a year to live and study with his grandfather, who was a well-known speech expert. With this background, it is hardly surprising that Bell became interested in the study of sound and the mechanics of speech.

When he returned to Edinburgh, Bell's first job was teaching elocution (the art of public speaking). He also assisted his father in his research and taught at a school for the deaf, using methods developed by his father. In 1870, Bell's family moved first to Brantford, Ontario, and then to Boston, Massachusetts. In Boston, Bell taught at the Boston School for the Deaf. In 1871, he opened his own school for the deaf and two years later, he became professor of speech and vocal physiology at Boston University. He also tutored private pupils who were deaf.

While teaching, Bell began experimenting with devices that would help deaf people hear and speak. His work ultimately led to the development of the first telephone in 1876. In 1877, Bell and his partners formed the Bell Telephone Company. His partners included his assistant, Thomas Watson (1854–1934), and the fathers of two of his deaf students. Bell later married one of these students, Mabel Hubbard.

Alexander Graham Bell. (Reproduced by permission of AT&T Bell Laboratories.)

As the telephone became a national phenomenon, Bell became a very wealthy man. He was able to build his own laboratory on Cape Breton Island, Nova Scotia, where he continued his research and invented a seemingly endless list of devices. Bell also became active in the science community. He was one of the first members of the National Geographic Society and served as the group's second president. He helped change the organization's magazine from a dry and relatively unpopular scientific journal into one of the most widely-read magazines in the world.

Bell continued to teach the deaf throughout his life. He also crusaded tirelessly for deaf rights and helped organize schools to integrate the deaf into society. In 1890, he established the Alexander Graham Bell Association for the Deaf. He became a good friend of Helen Keller (1880–1968) and was responsible for introducing her to Annie Sullivan, the woman who became her teacher.

Bell died in Nova Scotia on August 2, 1922. When he was buried, telephone service in the United States was stopped for one minute to honor the man who had contributed so much during his lifetime.

Technology and Invention

BIOGRAPHIES

■ KARL FRIEDRICH BENZ (1844–1929)

Karl Friedrich Benz was a German engineer, known for building the first modern gasoline-powered automobile.

Karl Benz was born in Karlsruhe, Germany, on November 26, 1844. His father, who was one of the first locomotive engineers, died when Benz was only two years old. Benz attended the local secondary school, where he studied chemistry and became interested in the popular new field of photography. He was especially intrigued by the mechanical sciences. In his autobiography, Benz wrote about his fascination with "the marvelous language that gear wheels talk as they mesh with one another."

At the age of sixteen, Benz entered a polytechnic school. There he studied steam engines and worked on designs to improve their operation. After graduation, Benz took a job in a machine shop. In 1871, he and partner August Ritter opened a factory to produce tin-working machinery. Benz's timing, however, was poor. The German economy was falling apart, and his business failed.

Karl Benz. (Reproduced by permission of AP/Wide World Photos.)

Benz's business troubles did not stop him from developing new steam engine designs. Soon he was building and selling a steam engine for use in stationary (non-moving) machines. In his spare time, he continued to work on his dream: a self-propelled vehicle that would operate by an internal combustion engine. According to Benz, this would be a much better alternative to the clumsy steam-powered vehicles that were being built.

By 1878, Benz had invented a two-stroke internal combustion engine. The engine worked so well and attracted so much attention that Benz

Technology and Invention

BIOGRAPHIES

opened up his own business in 1883, called Benz and Company. Benz's work benefitted greatly from another invention produced at about this time, the four-stroke internal combustion engine, developed by German engineers Nikolaus August Otto (1832–1891) and Eugen Langen (1833–1895). Building on the work of Otto and Langen, Benz had all of the elements needed to create the first true automobile.

In 1885, Benz and Company produced the first gasoline-powered motor vehicle. It was a three-wheeled, horseshoe-shaped conveyance that resembled a giant baby carriage. The car traveled at speeds of up to three miles per hour. Benz made his first commercial sale in 1887 in Paris, and within one year he had hired fifty men to help produce his three-wheeled motorcar.

Over the next decade, Benz continued to make improvements in the design of both the engine and the chassis (body). By 1893, he had successfully developed a four-wheel model called the Victoria. It was capable of traveling up to fifteen miles per hour. By 1902, Benz was producing a variety of models of various size and shape.

Benz left the company he founded in 1903 after years of dispute with the corporation's board of directors. He returned to the company briefly in 1904, but soon retired to Ladenburg, Germany, where he died in 1929. In the years following Benz's death, the company continued to grow and prosper. In 1926, it merged with the German firm of Daimler, to form one of the most famous automobile companies in the world, Daimler-Benz.

Henry Bessemer. (Reproduced by permission of UPI/Corbis-Bettmann.)

■ HENRY BESSEMER (1813–1898)

Henry Bessemer was an English inventor who developed an efficient, inexpensive method for producing steel. Because of Bessemer's invention, steel went from being a nearly precious metal to one that could be used for many everyday purposes.

Bessemer was born in Charlton, Hertfordshire, England, on January 19, 1813, the son of an engineer. From an early age, he was interested in mechanical devices and chemical processes. Because of his father' business, Bessemer visited many metal foundries and became familiar with all aspects of the metal industry. He learned about melting furnaces and how to combine various metals.

Throughout his life, Bessemer received over one hundred patents for his various inventions. One of his first inventions was a device for stamping deeds that saved the British government more than 100,000 pounds per year. Bessemer's most famous invention by far was the blast furnace, a device for making steel. Bessemer got his idea for a blast furnace from an earlier invention, a new type of cannon ball. The cannon ball he designed spun as it traveled through the air. The spinning action allowed the cannon ball to travel farther and more accurately. The cannon ball worked well, but cannons at the time were not strong enough to fire the new cannon ball.

The problem was that cannons were made from cast iron. Cast iron is a form of iron that contains relatively large amounts of carbon. Cast iron is very hard, but it breaks very easily. The only substitute available for cast iron at the time was wrought iron, which is nearly pure iron. Wrought iron was not suitable for making cannons (or practically anything else) because it was too soft.

The ideal material was steel, which has less carbon than cast iron, but more carbon than wrought iron. It combines the hardness and strength of cast iron with the durability of wrought iron. No one, however, had found an inexpensive and effective way to make steel with just the right amount of carbon—until Henry Bessemer.

Bessemer's idea was to blast cool air through molten cast iron. According to Bessemer, the oxygen in the air would burn away carbon in the cast iron, producing a durable type of steel. No one believed that Bessemer's process would work. Other inventors warned him that blowing cold air through molten cast iron would cool the iron too quickly and make it become solid.

The Bessemer steel process, the first inexpensive and effective way to make steel. (Courtesy of the Library of Congress.)

Technology and Invention

BIOGRAPHIES

The Bessemer process proved successful, however, and became the most important invention in the steel industry. Steel could now be mass produced and used to build such things as railroads and bridges. As the steel industry grew, it helped other industries, such as the automobile industry, grow into worldwide enterprises.

Bessemer continued to experiment with processes and inventions until he was over seventy years old. Among his later inventions were the first solar furnace and several diamond polishing machines. For his many accomplishments, Bessemer was asked to join Great Britain's prestigious Royal Society in 1877. In 1879, he was knighted by the British government. He died in London on March 15, 1898.

■ AUGUSTA ADA BYRON, COUNTESS OF LOVELACE (1815–1852)

Augusta Ada Byron was an English mathematician who is known for her contributions to the early field of computer science. She is credited with writing the first computer program, and her 1843 paper describing Charles Babbage's analytical engine showed an understanding of technology that was ahead of her time.

Byron was born in London in 1815, the daughter of Anne Isabella Milbanke, a wealthy heiress and amateur mathematician, and the famous poet George Gordon, Lord Byron (1788–1824). Byron's parents separated soon after her birth. She was raised by her mother, who insisted on a formal education for her daughter, even though it was uncommon for girls at the time. In addition to drawing, music, and French, Byron's tutors instructed her in mathematics and astronomy. Early on, Byron showed an aptitude for mathematics. By the age of five, she could add six rows of numbers.

Augusta Ada Byron. (Reproduced by permission of the Doris Langley Moore Collection.)

It was not long before Byron outgrew her tutors. She continued to study mathematics on her own and corresponded with mathematician Mary Somerville (1780–1872), who was the first woman elected to Great Britain's prestigious Royal Society. Unfortunately, because of the times, Byron was not permitted to attend a university, where she could continue her formal education.

In 1833, Byron met mathematician Charles Babbage (1792–1871). Babbage designed several types of machines to perform complex mathematical calculations. He also envisioned a machine, called an analytical engine, that could be programmed to perform calculations. Babbage's ideas would eventually lead to the development of the first computers. Limited by nineteenth-century technology, however, Babbage never actually built his machines.

Byron was intrigued by Babbage's vision and a life-long friendship formed between the two. In 1842, Babbage asked Byron to translate an article about his analytical engine from French to English. He also suggested that she add her own ideas and notes to the translation. In 1843, the translation and notes titled, "The Sketch of the Analytical Engine Invented by Charles Babbage," was published. Byron's paper was three times as long as the original article.

In her paper, Byron provided a clear mechanical explanation of how Babbage's engine worked. She also insisted that machines could not function without a programmer. Byron sketched out a series of operational instructions and created the world's first computer program.

At the time, almost no one knew that Byron had published the article. Since it was not thought proper for a woman to write such a technical paper, she signed only her initials to the translation. Twenty years after she died, Byron was finally acknowledged as the author of the article.

Byron's work was largely lost until 1953, when her notes were rediscovered and reexamined. By the 1970s, mathematicians and computer scientists acknowledged that Byron's work was exceptional for the time and clearly pointed to future technological possibilities. In the mid-1970s, the U.S. Department of Defense decided to create a single computer language so that all its various operations would be standard. In 1979, the department honored Byron's contribution to computer science by naming their high-level computer language ADA.

Technology and Invention

BIOGRAPHIES

■ THOMAS ALVA EDISON (1847–1931)

Thomas Edison is possibly the most famous inventor in history. Throughout his life, he patented more than one thousand inventions, a record number that has never been matched. His most famous inventions include the phonograph, the motion picture camera, and the incandescent light bulb.

Edison was born in Milan, Ohio, in 1847, but grew up in Port Huron, Michigan. His formal education consisted of three months at a public school when he was eight years old. His teachers dismissed him as a slow

Technology and Invention

BIOGRAPHIES

student, and he was taught at home by his mother. Edison was an avid reader and became interested in science, especially chemistry and electricity. As a child, he built a makeshift laboratory in his family's basement.

When he was twelve years old, Edison took a job selling candy and newspapers on a railroad. His experience with railroads led him to become an apprentice telegrapher. After the Civil War (1861–1865), Edison turned from telegraph operator to inventor. By 1869, he had made various improvements to the telegraph and was determined to pursue inventing as a full-time career. During this time, he continued to study on his own, reading the works of English physicist and electricity pioneer **Michael Faraday** (1791–1867; see biography in Physical Science chapter).

In 1876, Edison established a huge research and development laboratory in Menlo Park, New Jersey. As a goal, he predicted that he would create one new invention every ten days. During one period, he managed to patent a new invention every five days. Because of his amazing productivity, Edison was nicknamed "the wizard of Menlo Park."

Thomas Alva Edison in his laboratory. (Courtesy of the Library of Congress.)

Technology and Invention

BIOGRAPHIES

Edison and his team worked in Menlo Park until 1886. It was there that Edison improved on the telephone and created what some say was his most original invention, the phonograph. Invented in 1877, Edison's first phonograph was a hand-cranked machine that turned a tin-foil covered cylinder. A needle traced a groove on the cylinder.

At the same time, Edison was absorbed with the idea of creating a safe, inexpensive electric light that would replace gaslight. On New Year's Eve in 1879, he introduced the first incandescent electric light system by lighting up his laboratories and several streets in Menlo Park. By 1881, Edison had developed the first central power station and was supplying electric power to over eighty customers in New York.

By the 1880s, Edison had made millions of dollars from his inventions and, in 1887, he moved his operation to a much larger laboratory in West Orange, New Jersey. Just as he had in Menlo Park, Edison gathered together a group of top researchers, scientists, craftsmen, and businessmen to work on ambitious, technical advancements. In West Orange, one of Edison's teams invented the first movie camera, called the kinetograph, and the Kinetoscope, a device to show moving pictures.

Edison continued to develop many more projects, including a storage battery, electric railroads, and an improved method of manufacturing concrete. In 1914, a fire destroyed most of his West Orange complex, and by this time, Edison was beginning to finally slow down a little. Although opposed to war, he did help with U.S. naval research on torpedoes and periscopes during World War I (1914–18). In 1931, Edison died in West Orange at the age of eighty-four.

■ MARGARET E. KNIGHT (1838–1914)

Margaret Knight was an American inventor who developed so many devices she was nicknamed "Lady Edison." She is primarily known for inventing the square-bottomed paper bag still used in almost every grocery store.

Knight was born in York, Maine, on February 14, 1838. She was interested in tools and making things from a very early age. She once said that "the only things I wanted were a jack knife, a gimlet [a tool for boring holes], and pieces of wood." Knight attended local primary and secondary schools, but had no education beyond that point.

When she was twelve years old, Knight was inspired to create her first invention after visiting a cotton mill where her brothers worked. During her visit, she saw one of the mechanical looms break down. When it did, one of the loom's steel-tipped shuttles broke loose, flew across the room, and hit a worker. Knight invented a device to prevent such accidents. If a shuttle broke loose, the device shut the loom down automatically.

As an adult, Knight went to work in a shop making paper bags in Springfield, Massachusetts. She quickly saw problems with the bags she was making. They were shaped like an envelope and were not very strong. She found a way to mass produce paper bags with a flat bottom, like the ones in use today. These bags were not only stronger than the older bags, but they could stand up on their own. Knight received a patent for her invention in 1870.

During the 1880s, Knight turned out one new invention after another. She made a new kind of clasp for holding robes, a dress and skirt shield, a window sash and frame, and a skewer to hold meat for cooking. She then began to focus on the shoe-making business. She designed a number of machines that cut and sewed shoe leather more efficiently than could be done by hand. She earned six patents for her inventions in this field.

In the early 1900s, Knight found another new area of interest: motors and engines. She developed a number of new parts that made such machines work more efficiently.

By the time of her death in 1914, Knight had earned twenty-seven patents and made approximately ninety inventions. She is remembered as an extraordinary American inventor who, despite a lack of formal education, made great strides in all types of industry.

■ CYRUS HALL MCCORMICK (1809–1884)

American inventor Cyrus McCormick invented the mechanical reaper, which would have an enormous effect on agricultural life in the United States and abroad. The mechanical reaper not only made farm work faster, it prompted the development of other farm machinery, made it easier for pioneers to open up the American West, and helped farmers produce enough food for an ever growing population.

McCormick was born in Rockbridge County, Virginia, in 1809. His father was an inventor who patented several farm devices, including a thresher. His dream was to build a mechanical reaper, but he was never successful. Young McCormick had very little formal schooling, but he showed a talent for mechanics as he worked in his father's business.

In 1831, McCormick took on his father's dream and built the first horse-drawn reaper for harvesting wheat. It could isolate the grains on the stalks, hold them in position for cutting with a blade, and then collect the grains. Using McCormick's reaper, a farmer could cut the same amount of wheat in one day that had previously taken several men more than two weeks. Although the reaper was modified over time, the basic design of McCormick's reaper remains the same today.

By 1837, McCormick was building reapers at his family's blacksmith shop in Walnut Grove, Virginia. To make more money, he sold his patent to several manufacturers in New York and Ohio. In 1847, McCormick had made enough to build his own factory in Chicago, Illinois, where he produced eight hundred reapers his first year.

McCormick continued to improve on his machine and he exhibited it throughout the United States. Salesmen visited farms in rural America to demonstrate the amazing reaper, offering payment plans to help farmers afford new machinery. In 1851, McCormick began selling his reaper in Europe, and soon he was the largest manufacturer of farm machinery in the world. Because of his enormous success, McCormick became a prominent business leader in the United States, who had quite a bit of influence with government officials and members of Congress.

In 1871, during the great Chicago fire, McCormick's factory burned to the ground. McCormick rebuilt his company and, in the process, rebuilt much of Chicago. For that reason, many Chicago landmarks are named after him. During the 1880s, McCormick was selling 50,000 reapers per year and his company employed over 1,000 workers.

At the time of his death in 1884, McCormick had produced six million harvesters. Because of his investments in railroad and mining stock, he left behind a sizable fortune. In 1902, the McCormick Harvester Company merged with Deering Harvesting Company and several other farm machinery producers to become the International Harvester Corporation.

Technology and Invention

BIOGRAPHIES

Cyrus Hall McCormick. (Courtesy of the Library of Congress.)

■ NORBERT RILLIEUX (1806–1894)

Norbert Rillieux was an African-American inventor and engineer who developed machinery used in sugar refining. His inventions greatly

Technology and Invention

BIOGRAPHIES

expanded the sugar industry and helped ease the work that was done by American slaves.

Rillieux was born in New Orleans, Louisiana, on March 17, 1806. His mother was of mixed race and his father was a white plantation owner. Rillieux showed an early gift for learning and was sent to France for his college education. There he attended the l'École Centrale in Paris. After receiving his degree in engineering, he became a teacher at the school.

While in France, Rillieux became interested in improving the method used to refine sugar. As a young man, he had watched how sugar was refined on Louisiana plantations. At the time, refining sugar was a laborious, and sometimes dangerous, task that was performed primarily by slaves. Slaves tended the fires used to heat sugar cane juice in large, open vats. As the liquid boiled down, it was transferred to smaller vats by slaves, who scooped the hot liquid with ladles. Eventually, the liquid boiled away completely, leaving behind sugar crystals. Most of the sugar was brown and of poor quality. Pure white sugar was very expensive to make and could be purchased only by the rich.

Rillieux invented a safer and more efficient device for making sugar crystals. In his device, sugar cane juice was heated in a series of closed vats. Some of the air from the vats was pumped out, which allowed the sugar cane juice to boil at a lower temperature. Steam and heated juice were pumped from one vat to the next through pipes. Steam from the first vat helped the liquid in the second vat to boil, and so on, through the entire series of containers. As a result, white sugar crystals formed more easily and were of good quality. In addition, because Rillieux's device recycled the heat used to boil the sugar cane juice, the process became much less expensive.

Rillieux left Paris in 1834 and returned to New Orleans, where he built and tested his evaporator. Although his first tests were unsuccessful, Rillieux was later able to make the device work. He obtained a patent for his invention in 1846. Before long, thousands of Rillieux's evaporators were in use in both the United States and the Caribbean.

After the success of his evaporator, Rillieux turned his attention to the design of a sewage and drainage system for New Orleans. The city, which lies at the mouth of the Mississippi River, had long been plagued by serious sanitation problems. His plan was rejected by city officials, although one very similar to it was later adopted.

In the 1850s, Rillieux left New Orleans because of increased restrictions on free African Americans. He returned to Paris where he continued to work as an engineer. He also developed an interest in Egyptian history and worked on deciphering Egyptian hieroglyphics for almost twenty years. He died in Paris on October 8, 1894.

Technology and Invention

BIOGRAPHIES

◼ ISAAC MERRIT SINGER (1811–1875)

American inventor and businessman Isaac Singer did not actually invent the sewing machine, but he did develop the first practical machine to be commercially successful. Singer also invented a number of other useful items, including a rock-drilling machine and a metal- and wood-carving machine. In addition, he was a clever businessman who introduced new ideas into sales and marketing.

Singer was born in Pittstown, New York, on October 27, 1811. He had relatively little formal education and became an apprentice machinist at a young age. He invented his rock-drilling machine when he was twenty-eight years old; he invented a metal/wood-carving machine ten years later. Singer's first experience with sewing machines came in 1850, when he was asked to inspect a machine that had been brought in for repair. Singer repaired the machine quickly, and saw a number of improvements that could be made to the existing design. Less than two weeks later, he had built his own version of the mechanical sewing machine. By 1851, Singer had patented his invention.

At that point, Singer became involved in a series of legal problems. He was, after all, not the first person to invent a mechanical sewing machine. That honor goes to French inventor Barthélemy Thimonnier (1793–1859). In 1830, Thimonnier had patented a machine for the production of army uniforms. In the United States, Elias Howe (1819–1867) had produced and patented an improved version of Thimonnier's machine as early as 1845. When Howe heard of Singer's invention, he sued to prevent Singer from marketing his sewing machine.

The lawsuit dragged on for some time, and Howe eventually won his case. But in the meantime, Singer formed his own company and began manufacturing various versions of his invention. By 1860, the Singer Manufacturing Company was the world's large manufacturer of sewing machines.

Singer continued to improve the sewing machine for the next few years. Many of his improvements involved the way a sewing machine was powered. The first machines were operated by means of a hand crank, which left tailors only one hand free to sew. Singer solved that problem by inventing a foot treadle. Tailors could then operate a machine with their feet, leaving both hands free for sewing.

Singer was not only an inventor, he was a sharp businessman who had very specific ideas about how his machines should be marketed and sold. He

Technology and Invention

BRIEF BIOGRAPHIES

was one of the first manufacturers to use the idea of installment paying for purchasing an item. This method allowed people to buy expensive equipment over time that they might not otherwise have been able to afford.

Singer made a fortune from his sewing machines and was known for his lavish lifestyle. He retired from the sewing machine business in 1863 to travel through Europe before retiring to England. He died in 1875 at his home in Torquay. Singer's company survived for another 113 years. During that time, the company continued to make improvements in the sewing machine, introducing the first electric-powered machine in 1885 and the first mass-produced machine in 1910.

BRIEF BIOGRAPHIES

▲ CLÉMENT ADER (1841–1925)

Ader was a French inventor who experimented with aircraft beginning in 1882. He designed and built several machines with bat-shaped wings. He attempted to fly a steam-powered aircraft called the Eole in 1890. The aircraft took off but failed to remain airborne. Ader is credited with inventing the French term for airplane, *avion*.

▲ ALEXANDER BAIN (1810–1903)

Bain was a Scottish clockmaker and inventor who designed the first method for transmitting images over distances. His invention was the earliest form of the facsimile, or "fax," machine used so widely today. Bain also made other useful inventions, including an electric clock and a devices for controlling the operation of steam railway engines.

▲ LOUIS BRAILLE (1809–1852)

Braille was a French educator who developed a system of printing and writing for use by blind people. The system consists of a six dots that can be arranged in various combinations to spell out letters and numbers. The system is now known by his name, the Braille system. Although he lost his sight at the age of three, Braille become an accomplished musician and inventor. In addition to his system of printing and writing, Braille developed a separate code for reading and writing music and mathematics.

▲ ISAMBARD KINGDOM BRUNEL (1806–1859)

Isambard Brunel (son of Marc Isambard Brunel) was one of England's greatest eighteenth century inventors. He designed and supervised the construction of the rail line running from London to Bristol, commonly known as the Great Western Railway. His innovations in bridges, tunnels, and track helped reform England's locomotive industry. Brunel was also a pioneer in steam navigation. He designed three of the world's great steamships—the Great Western, (1838), the Great Britain (1845), and the Great Eastern (1858), the largest steam vessel of its time.

Technology and Invention

BRIEF BIOGRAPHIES

▲ MARC ISAMBARD BRUNEL (1769–1849)

Marc Brunel (father of Isambard Kingdom Brunel) was a clever man who created many new inventions, including machines for sawing and bending wood, making boots, knitting stockings, manufacturing nails, and making copies of drawings. His most famous invention was a machine for digging tunnels under a river. Between 1825 and 1843 he was in charge of building a tunnel under the Thames River. The tunnel is now part of the London subway system.

▲ WILLIAM SEWARD BURROUGHS (1855–1898)

Burroughs was an American inventor who designed and built one of the first modern calculators. He became interested in inventing at an early age while tinkering in his father's machine shop. At the age of fifteen, he invented a device that could calculate the sum of two numbers. He continued working on his machine for many years until he produced a calculator that could print its results on paper. His invention proved to be a great success, although not in Burroughs's lifetime.

▲ FERDINAND CARR (1824–1894)

Carr was a French engineer who invented the first mechanical refrigerator. The machine operated by using gaseous ammonia. Carr, also conducted research in the field of electricity. He invented a device to regulate electric lights and a mechanism for producing high voltages.

▲ SAMUEL COLT (1814–1862)

Colt was an American inventor who designed and built the first revolver, also known as a "six-shooter." His guns were not very popular when he first started producing them because the country was at peace. Once the Civil War broke out, however, they became the most widely used weapon in the war. The Colt Industry factory made extensive use of assembly lines and interchangeable parts, both major innovations in the manufacturing industry.

▲ JOHN FREDERIC DANIELL (1790–1845)

Daniell was an English chemist best known for his invention of the Daniell cell. The Daniell cell was the most advanced kind of battery avail-

Technology and Invention

BRIEF BIOGRAPHIES

able at the time. Daniell also invented the hygrometer, an instrument for measuring relative humidity and dew point. Daniell was also active as an educator, philosopher, and writer.

▲ RUDOLF DIESEL (1858–1913)

Diesel was a German engineer who invented a type of internal combustion engine that now carries his name. Diesel first used powdered coal as a fuel for the engine, but later switched to kerosene. The diesel engine was not very popular with automobiles at first, but did find many applications in the shipping and locomotive industries.

▲ GEORGE EASTMAN (1854–1932)

Eastman was an American inventor who devised a simplified method of photography that could be used by almost anyone. Prior to his invention, a large and cumbersome set of equipment was needed to take a photograph. Eastman eventually obtained patents for the first simple Kodak box camera (1888), celluloid film rolls (1889), and the even simpler Kodak Brownie camera (1900) that dominated the amateur photography market for decades.

▲ ALEXANDRE-GUSTAVE EIFFEL (1832–1923)

Eiffel was a French engineer who developed new designs for steel bridges and other buildings. In 1885, he designed the internal metal structure for the Statue of Liberty. Two years later, he supervised the construction of the 900-foot tall tower that now bears his name. Towards the end of his life, he became interested in the field of aerodynamics.

▲ JOHN ERICSSON (1803–1889)

Ericsson was a Swedish-born American engineer who invented the screw propeller as a method for driving steamships. The screw propeller replaced the paddle wheel, which was large and cumbersome. In 1861, Ericsson oversaw the construction of the steamship Monitor, the first entirely ironclad steamship.

▲ HEINRICH GEISSLER (1815–1879)

Geissler was a German inventor who developed the Geissler tube and the Geissler mercury pump. A Geissler tube is a sealed glass tube containing a near-vacuum through which an electrical current is transmitted. It is the forerunner of the cathode ray tube, used today in televisions and computer monitors.

Technology and Invention

BRIEF BIOGRAPHIES

▲ HENRI GIFFARD (1825–1882)

Giffard was a French engineer who built the first lighter-than-air aircraft. He attached a 3-horsepower motor driven by a steam engine to a 144-foot-long balloon. The ship was the forerunner of the modern dirigible perfected by German inventor Ferdinand von Zeppelin.

▲ ELIAS HOWE (1819–1867)

Howe was an American inventor who patented the first sewing machine in the United States. However, Howe was unable to interest manufacturers in his machine. He eventually sold his patent to William Thomas in 1847. Unlike Howe, Thomas made a fortune on the machine. It was not until late in his life that Howe finally received income from his original invention. He won a law suit that brought him royalties on all sewing machines manufactured in the United States between 1854 and 1867.

▲ SAMUEL PIERPONT LANGLEY (1834–1906)

Langley was an American astronomer who did pioneer research in heavier-than-air flight. He constructed aircraft that were able to fly, but, unfortunately, unable to carry humans. His last unsuccessful flight was attempted just shortly before the Wright brothers made the world's first successful flight in 1903. Langley also had a distinguished scientific career and served as the Secretary of the Smithsonian Institution from 1887 until his death in 1906.

▲ JEAN-JOSEPH-ÉTIENNE LENOIR (1822–1900)

Lenoir was a Belgian-French inventor who developed the first useable internal combustion engine. His first design made use of a coal gas/air mixture to operate the engine. He later modified the device to operate on liquid fuel. He used the engine to drive a land vehicle in 1862 and a boat in 1886.

▲ HIRAM STEVENS MAXIM (1840–1916)

Maxim was an American-born British inventor who invented the first fully automatic machine gun. This weapon was first used by the British in their conquest of Africa and Asia. The horrors of its effectiveness were most clearly illustrated during World War I. Maxim's many other inventions included a steam-powered flying machine, an electric pressure regulator, smokeless gunpowder, a pneumatic gun, carbon filaments for light bulbs, vacuum pumps, gas motors, and an automatic, steam-powered water pump.

▲ JOHN LOUDON MCADAM (1756–1836)

McAdam was a Scottish engineer often called the Father of Modern Road Building. He used crush rocks as the basis for his roads, a practice on which modern road building is still based. This method is called "macadamizing" is named in his honor.

▲ WILLIAM MURDOCK (1754–1839)

Murdock was a Scottish-born English inventor who pioneered the use of coal gas for lighting. His system was the first new system of lighting in the modern age. His first large project was the installation of exterior lighting at the Boulton and Watts factory in the Soho district of London. By 1807, his lighting system was being used on many London streets.

▲ ELISHA GRAVES OTIS (1811–1861)

Otis was an American inventor who designed the first "safety hoist" elevators. These elevators are designed not to fall even if the cable that holds them breaks. The first passenger elevator using this system was installed in a New York store on 23 March 1857. The elevator company started by Otis to sell his invention is still in business.

▲ NIKOLAUS AUGUST OTTO (1832–1891)

Otto was a German inventor who built the first four-stroke internal combustion engine. The term four-stroke is used to describe the four stages through which the engine goes: intake, compression, combustion/power, and exhaust. The four-stroke cycle is still known as the Otto cycle in his honor. It is the basis for most automobile and airplane engines today.

▲ ALEXANDER PARKES (1813–1890)

Parkes was an English chemist who discovered the first plastic. He made this plastic by dissolving cellulose nitrate in alcohol and ether in the presence of camphor. When the liquids in the mixture evaporated, a hard solid formed. That material was later named parkesine. Parkes had no success in selling this product. But it later became popular in the production of ornaments, knife handles, and fishing reels.

▲ GEORGE PULLMAN (1831–1897)

Pullman was an American inventor who built the first railroad sleeping car. In 1867, he founded his own company to build and sell his cars. He later developed the first dining car (in 1868) and the first chair car (in 1875).

▲ ERNEST WERNER VON SIEMENS (1816–1892)

Siemens was a German-British engineer and inventor who made important contributions to the electrical industries of Germany, Great Britain, and

Russia. He also worked on the early development of the electric locomotive, designed a ship for the laying of a transatlantic cable, developed the telegraph system in his native Prussia, invented the dial telegraph, and made important improvements in the electric generator. The company he founded remains in existence today.

Technology and Invention
BRIEF BIOGRAPHIES

▲ GEORGE STEPHENSON (1781–1841)

Stephenson was an English inventor who built some of the world's first railroad engines. He had little formal education, but had a genius for inventing new devices. In 1841, he introduced his first steam locomotive. It was able to pull 30 tons of material more rapidly than a horse-drawn cart. His most famous engine was the "Rocket," which could travel at the unheard of speed of 60 kilometers per hour (36 miles per hour).

▲ JOSEPH WILSON SWAN (1828–1914)

Swan was an English chemist and physicist who invented the electric light bulb at about the same time as Thomas Edison. Swan also invented the carbon process of printing (1864), the dry photographic plate (1871) and bromide photographic paper (1879). In 1883, he invented a process for squeezing nitrocellulose through small holes to produce fibers. He was unable to market this invention, although it later led to the development of the first artificial fibers.

▲ WILLIAM HENRY FOX TALBOT (1800–1877)

Talbot was an English inventor who developed one of the earliest forms of photography. His system was similar to that invented by Louis Daguerre. But Talbot also introduced the use of negatives, from which positive prints could be made. He published the first book illustrated with photographs, Pencil of Nature, in 1844.

▲ FREDERICK W. TAYLOR (1856–1915)

Taylor was an American engineer who is most famous for developing a system of mass production known as scientific management. He used time studies, job analysis, and other techniques to find the most efficient way to do each job on a production line. He then developed work policies and practices that would reduce wasted time and effort by workers. Many workers opposed his methods, but they became very popular among business owners who saw scientific management as a way of increasing their profits.

▲ NIKOLA TESLA (1856–1943)

Tesla was a Croatian-born American electrical engineer who invented a number of important electrical devices. Among these devices was a trans-

Technology and Invention

RESEARCH AND ACTIVITY IDEAS

former that made it possible to send electrical currents over very long wires. He also invented a motor that would run on alternating current (AC). The unit of magnetic flux in the metric system is now named the tesla in his honor.

▲ GRANVILLE T. WOODS (1856–1910)

Woods was an American engineer and inventor who received approximately sixty patents. He is considered the most productive African American inventor of the nineteenth century. Most of his inventions were railroad related. However, he also patented an egg incubator, a telephone transmitter, and an electric car.

RESEARCH AND ACTIVITY IDEAS

- It is often said that science and technology are international activities. People from all around the world often contribute to the development of a single new idea. Select any one invention from the nineteenth century to see how true this claim is. Make a chart, table, diagram, or illustration showing how men and women from more than one country had a part in the development of the invention.

- The inventions produced in the nineteenth century did not develop in isolation from each other. One invention often contribution to the development of one or more other inventions. For example, Henry Bessemer's method of making inexpensive steel made possible many other developments. Make a "Technology Tree" that shows how two or more inventions were related to each other.

- Historians point out that science and technology are mutually dependent on each other. For example, a scientific discovery will often lead to one or more inventions. And a new invention may make scientific discoveries possible. Find one example of the way this mutual dependence works in each direction. Write a short article describing your findings.

- Most inventions that have changed the modern world were made in the Western world, especially western Europe and the United States. Develop a hypothesis as to why this fact is true. Then see what information you can find to support or reject your hypothesis. Prepare a reading list of books and articles that you consulted on this topic.

● Nineteenth-century inventions made possible a host of new products and materials. But they also made some important changes in the way people live. They altered housing, work, educational, and other patterns in the Western world. Select any one invention and show how it brought about such changes. Write a story about a family living in the nineteenth century whose lives might have been changed by the invention you selected.

FOR MORE INFORMATION

Books

Austrian, Geoffrey D. M. *Herman Hollerith: Forgotten Giant of Information Processing.* New York: Columbia University Press, 1984.

Baum, Joan. *The Calculating Passion of Ada Byron.* Hampden, CT: Archon Press, 1986.

Barger, M. Susan, and William B. White. *The Daguerreotype: Nineteenth-Century Technology and Modern Science.* Baltimore: Johns Hopkins Press, 2000.

Brooks, John. *Telephone: The First Hundred Years.* New York: Harper & Row, 1976.

Brown, G. I. *The Big Bang: A History of Explosives.* Stroud, UK: Sutton Publishing, 1998.

Cardwell, Donald. *The Norton History of Technology.* New York: W. W. Norton & Company, 1994.

Carnegie, Andrew. *The Autobiography of Andrew Carnegie.* Boston: Northeastern University Press, 1990.

Davidson, Margaret. *The Story of Alexander Graham Bell: Inventor of the Telephone.* Milwaukee: Gareth Stevens, 1997.

Dickinson, H. W. *A Short History of the Steam Engine.* New York: MacMillan, 1939.

Fisher, Leonard Everett. *Alexander Graham Bell.* New York: Atheneum Press, 1999.

Giannetti, Louis, and Scott Eyman. *Flashback: A Brief History of Film.* Englewood Cliffs, NJ: Prentice-Hall, 1986.

Headrick, Daniel R. *The Tools of Empire: Technology and European Imperialism in the Nineteenth Century.* Oxford: Oxford University Press, 1981.

McLeod, Elizabeth. *Alexander Graham Bell: An Inventive Life.* Buffalo, NY: Kids Can Press, 1999.

Technology and Invention

FOR MORE INFORMATION

Mead, George. "Negro Scientist of Slavery Days." *Negro History Bulletin* (April 1957) 159–63.

Meltzer, Milton. *The Many Lives of Andrew Carnegie.* New York: Franklin Watts, 1997.

Moran, James. *Printing Presses: History and Development from the Fifteenth Century to Modern Times.* Berkeley: University of California Press, 1973.

Morus, Iwan Rhys. *Frankenstein's Children: Electricity, Exhibition, and Experiment in Early-Nineteenth-Century London.* Princeton, NJ: Princeton University Press, 1998.

Noonan, Jon. *Nineteenth-Century Inventors.* New York: Facts on File, Inc., 1992.

Payson, Abbott. *A History of Mechanical Inventions.* Cambridge, MA: Harvard University Press, 1954.

Peters, Tom F. *Building the Nineteenth Century.* Cambridge, MA: MIT Press, 1996.

Reid, James D. *The Telegraph in America: Its Founders, Promoters, and Noted Men.* New York: Arno Press, 1974.

Schivelbrusch, Wolfgang. *Disenchanted Night: The Industrialization of Light in the Nineteenth Century.* Berkeley: University of California Press, 1995.

Siegel, Beatrice. *The Steam Engine.* New York: Walker & Company, 1986.

Simon, Charnan. *Andrew Carnegie: Builder of Libraries.* Chicago: Childrens Press, 1998.

Smith, Geoffrey. *The Oxford History of World Cinema.* Oxford: Oxford University Press, 1996.

Stein, Dorothy. *Ada: A Life and a Legacy.* Cambridge, MA: MIT Press, 1985.

Stern, Ellen, and Emily Gwathmey. *Once upon a Telephone: An Illustrated Social History.* New York: Harcourt Brace, 1994.

Taylor, George Rogers. *The Transportation Revolution, 1815–1860.* New York: Armonck Press, 1968.

Toole, Betty Alexandra. *Ada: The Enchantress of Numbers.* Stewart, OH: Strawberry Press, 1992.

Wade, Mary Dodson. *Ada Byron Lovelace: The Lady and the Computer.* Englewood Cliffs, NY: Silver Burdett Press, 1994.

Wall, Joseph Frazier. *Andrew Carnegie.* Pittsburgh: University of Pittsburgh Press, 1969.

Wenden, D. J. *The Birth of the Movies.* New York: E. P. Dutton, 1975.

Web sites

The American Experience: The Telephone. [Online] http://www.pbs.org/wgbh/amex/telephone/index.html (accessed on February 26, 2001).

AWC Lovelace Award—Who Was Ada Lovelace? [Online] http://www.awc-hq.org/lovelace/whowas.htm (accessed on February 21, 2001).

Technology and Invention

FOR MORE INFORMATION

"Computers: History and Development." *Jones Telecommunications & Multimedia Encyclopedia.* [Online] http://www.digitalcentury.com/encyclo/update/comp_hd.html (accessed on February 21, 2001).

"Henry Bessemer, Man of Steel." *ExNet Online: Science and Technology.* [Online] http://www2.exnet.com/1995/09/27/science/science.html (accessed on February 21, 2001).

Herman Hollerith's Tabulating Machines. [Online] http://www.maxmon.com/1890ad.htm (accessed on February 21, 2001).

Herman Hollerith Inventor. [Online] http://inventors.about.com/science/inventors/library/inventors/blhollerith.htm?terms=herman+hollerith (accessed on February 21, 2001).

The Norbert Rillieux Home Page. [Online] http://www.norfacad.pvt.k12.va.us/project/rillieux/Rillieux.htm (accessed on February 21, 2001).

"Norbert Rillieux: Sugar Chemist and Inventor." *Faces of Science: African Americans in the Sciences.* [Online] http://www.princeton.edu/~mcbrown/display/rillieux.html (accessed on February 21, 2001).

Sewing Machine History. [Online] http://www.icsi.net/~pickens/msinger.shtml (accessed on February 21, 2001).

The Telegrapher Web Page. [Online] http://www.mindspring.com/~tjepsen/Teleg.html (accessed on February 26, 2001).

Index

Italic types indicates volume number; **boldface** type indicates main entries and their page numbers; **(ill.)** indicates illustrations.

A

Abel, Frederick *2:* 346
Acoustics *2:* 232, 233
Acquired characteristics, inheritance of *1:* 30
Acupuncture *1:* 85
ADA (computer language) *2:* 201, 359
Adams, John Couch *2:* 267, 285
Adaptation *1:* 99
Addison, Thomas *1:* 149
Ader, Clément *2:* 366
Adler, Alfred *1:* 137
African American inventors *2:* 363, 372
Agassiz, Jean Louis *2:* 261, **265–66**, 265 (ill.), 286
Agoraphobia *1:* 156
Agricultural sciences *1:* **16–21**
    Hatch Act *1:* 19
Aircraft *2:* 365, 369
Algebra, boolean. *See* Boolean algebra
Algebra, fundamental theorem of *2:* 191
Algebraic geometry *2:* 207
Alkaloid chemistry *1:* 72
Alternating current (AC) *2:* 325, 326, 372
Alternative medicine *1:* **84–89**
Aluminum *2:* 260
Amalgam *1:* 107, 108
American Asylum for the Deaf and Dumb *1:* 103
American Dental Association *1:* 109
American Institute of Homeopathy *1:* 85
American Mathematical Society *2:* 174
American Museum of Natural History *1:* 47
American Public Health Association *1:* 124
American Society of Dental Surgeons *1:* 109
Ampere, Andre Marie *2:* 285
Analgesia *1:* 87
Analytical engine *2:* 299, 357, 358
Analytical Society *2:* 183
Anaphase *1:* 15

xxxix

Index

Anderson, Elizabeth Garrett *1:* 135
Andral, Gabriel *1:* 149
Anesthesia *1:* 87, 90, 154
Anthrax *1:* 40, 44, 95, 125, 139
Anthropology *1:* 8
Antibodies *1:* 44
The Antiquity of Man *2:* 283
Antiseptics *1:* 87, 93
Appert, Nicolas *2:* 301, 341, **352–53**, 353 (ill.)
Arcueil circle *2:* 275
Argon *2:* 244
Aristotle *1:* 21, *2:* 176
Armat, Thomas *2:* 322
Armour, Philip *2:* 343
Arrhenius, Svante August *2:* 285
Arsphenamine *1:* 136
Art therapy *1:* 110
Aseptic techniques *1:* 82, 147
Assembly line *2:* 304, 307
Association of American Physicians *1:* 143
Astronomy
    Herschel, Caroline *2:* **280–82**
    mathematics of *2:* 195, 203
Atomic theory *2:* 213, 214, 271
    Henry, William *2:* 288
Atomic weight *2:* 227, 228
Atomic weights, table of *2:* 272
Atwater, Wilbur *1:* 20
Audubon, John *1:* **54–56**
Auer, Carl *2:* 285
Automobiles *2:* **303–8**
    assembly line *2:* 307
    gasoline powered *2:* 354
    internal combustion engine *2:* 300
Automorphic functions *2:* 200
Avogadro, Amedeo *2:* 229, 277
Ayrton, Bertha *2:* 286

B

Babbage, Charles *2:* **182–85**, 183 (ill.), 299, 358
    Byron, Augusta Ada *2:* 201, 359
    punched cards *2:* 338
Bacteria, classification of *1:* 68
Baer, Karl von *1:* 5
Bain, Alexander *1:* 149, 365
Baird, Spencer *1:* 56–58, 56 (ill.)
Ballistite *2:* 346
Bandages *1:* 114
Barium *2:* 258
Bartels, Johann *2:* 198
Barton, Clara *1:* 84, **128–29**
Bates, Henry *1:* 65, 67
Batteries *2:* 169
    Daniell cell *2:* 367
    voltaic pile *2:* 169, 226, 324
Battlefield medicine *1:* 113
Beaumont, William *1:* 81, **130–31**, 130 (ill.), 131 (ill.)
Becquerel, Antoine *2:* 237, 238–39, 269
Beddoes, Thomas *1:* 90
Behring, Emil von *1:* 135, 139
Beijerinck, Martinus *1:* 44, 67
Bell, Alexander Graham *2:* 224, 301, **316–20**, 317 (ill.)
Bell curve *2:* 180
Bell Telephone Company *2:* 319, 353
Beltrami, Eugenio *2:* 200
Benz, Karl *2:* 305, 306 (ill.), **355–56**, 354 (ill.)
Bernoulli, Jakob *2:* 180
Berthollet, Claude *2:* 275
Berzelius, Jöns *1:* 74, *2:* 228, 286
Bessel, Friedrich *2:* 215, **267–69**, 267 (ill.)
Bessemer, Henry *2:* **356–58**, 356 (ill.)

Index

Bessemer process 2: 356
Betti, Enrico 2: 201
Bichat, Marie François 1: 10, 150
Biggs, Hermann M. 1: 96
Billroth, Christian 1: 93, 150
Binary system 2: 177
Biot, Jean-Baptiste 2: 276
Birds of America 1: 54
Black bodies 2: 289
Blackwell, Elizabeth 1: 84, **132–35**, 132 (ill.), 157
Blackwell, Emily 1: 134
Blast furnace 2: 357
Blind
    printing and reading system for 1: 104, 2: 366
    schools and education 1: 104–5
Blood disorders 1: 149
Blood-letting 1: 40
Blumenbach, Johann 1: 4, **7–10**, 9 (ill.)
Body temperature 1: 156
Bohr, Niels 2: 226
Boiler 2: 304
Bolyai, Janos 2: 165, 197
Bolzano, Bernhard 2: 201
Bond, George 2: 286
Bond, William 2: 286
Boole, George 2: 166, 176, **185–86**
Boole, Mary 2: 201
Boolean algebra 2: 166, 176, 185–86
Borden, George 2: 343
Boron 2: 276
Botany, scientific 1: 13
Boucher de Crevecoeur de Perthes, Jacques 1: 49
Bouchon, Basile 2: 336
Boveri, Theodor 1: 67
Bowman, William 1: 150

Boyle, Robert 2: 243
Braidwood, Thomas 1: 103
Braille, Louis 1: 105, 2: 366
Braille system 1: 104, 2: 366
Bretonneau, Pierre-Fidele 1: 150
Bridges (dental) 1: 107
Brisseau-Mirbel, Charles 1: 13
British Museum 1: 47
Brongniart, Alexandre 1: 67
Brown, Robert 1: 14, 67
Bruce, David 1: 150
Brunel, Isambard 2: 366
Brunel, Marc 2: 367
Bubonic plague 1: 126, 139, 140
    cause of 1: 157
Buchner, Hans 1: 151
Bunsen, Robert 2: 286
Burbank, Luther 1: 20
Burdach, Karl 1: 3
Burroughs, William 2: 367
Byron, Augusta Ada 2: 201, **358–59**, 358 (ill.)

C

Calcium 2: 258
Calculus 2: 165, 166
    acoustics 2: 224
Calculus of vectors 2: 203
Camera 2: 322
Camera lucida 2: 293
Candolle, Augustin-Pyrame de 1: 68
Canning (foods) 2: 341, 342, 353
Cannizzaro, Stanislao 2: 229
Cantor, Georg 2: 166
Cantor, Moritz 2: 201
Caoutchouc 2: 349
Carbohydrates 1: 71
Carbolic acid 1: 87
Carlisle, Anthony 2: 256
Carnot, Sadi 2: 214, 278
Carr, Ferdinand 2: 367

Index

Cast iron 2: 357
Catastrophism 2: 252, 282
Cathode ray tube 2: 368
Cathode rays 1: 145
Cauchy, Augustin-Louis 2: 166, 186–87
Caventou, Joseph-Beinaime 1: 72
Cell division 1: 14
Cell theory 1: 5, **10–16**
    Brown, Robert 1: 67
    Mohl, Hugo von 1: 14
    Oken, Lorenz 1: 65
Cells 1: 11
Celluloid 2: 320, 322
Cephalic index 1: 99
Cerebellum 1: 69
Cerium 2: 285
Chadwick, Edwin 1: 122
Chappe, Claude 2: 313
Charcot, Jean Martin 1: 137
Charles, Jacques 2: 276
Charles' Law 2: 276
Charpentier, Johann von 2: 261, 286
Chasles, Michel 2: 202
The Chemical History of the Candle 1: 47, 2: 275
Chemical notation 2: 285
Chemistry, alkaloid 1: 72
Chiropractic medicine 1: 84, 85, 88
Cholera 1: 118, 122, 125
Chromatin 1: 14, 69
Chromosomes 1: 11, 67
Cinematographe 2: 322
Clark, Edward H. 1: 102
Clausius, Rudolf 2: 286
Clerc, Laurent 1: 103
Clifford, William 2: 202
Climate, ice ages and 2: 263
Cocaine 1: 154
Cochlea 2: 224, 233

Cohn, Ferdinand 1: 68
Colt revolver 2: 367
Colt, Samuel 2: 367
Comets 2: 287
Computers 2: 337, 358
    boolean logic and 2: 178
    punched cards 2: 338
Computers, early 2: 183
Condamine, Charles Marie de la 2: 349
Conditioned reflexes 1: 155
Conservation of energy, law of. *See* Law of conservation of energy
Consumption. *See* Tuberculosis
Copernicus, Nicholas 2: 268
Cordite 2: 345, 346
Coriolis, Gaspard 2: 205
Corporations, rise of 2: 334
Correlation 2: 182
Correns, Karl 1: 22
Cosmology 2: 252
Cournot, Antoine 2: 202
Craniology 1: 61
Creationism 1: 30, 50, 2: 251, 252
Crede, Karl 1: 151
Crelle, August 2: 174, 202
Crelle's Journal 2: 202
Cro-Magnon man 1: 49
Crowns (dental) 1: 107
Cugnot, Nicolas-Joseph 2: 304
Curie, Marie 2: 240, **269–71**, 270 (ill.)
Curie, Pierre 2: 240
Curves 2: 180
Cuvier, Georges 1: 51, 52 (ill.)
Cytology 1: 11, 22
Cytoplasm 1: 73

D

D'Ocagne, Philbert 2: 203
Daguerre, Louis 2: 302, 371
Daimler, Gottlieb 2: 306

Index

Dalton, John 2: 214, 228, **271–73**, 272 (ill.)
Daniell cell 2: 367
Daniell, John 2: 367
Darwin, Charles 1: 6, 10, **29–36**, 29 (ill.)
    Huxley, Thomas 1: 60
    ice ages 2: 266
    Lyell, Charles 2: 283
    population studies 1: 25
    theory of evolution 1: 29–36, 48
    uniformitarianism 2: 254
Davis, William 2: 287
Davy, Humphry 1: 18, 2: 257–58, 275
    Faraday, Michael 2: 273
De Magnete 2: 221
De Morgan, Augustus 2: 202
De Mortillet, Louis-Laurent-Marie Gabriel 1: 71
Deaf, education of 1: 103–4
Debridement 1: 114
DeBries, Hugo 1: 22
Deductive reasoning 2: 176, 177
Democritus 2: 272
Dental drill 1: 108
Dentistry, advances in 1: **106–9**
Dentures 1: 108 (ill.)
Descartes, René 2: 213
The Descent of Man 1: 35
Dewar flask 2: 287
Dewar, James 2: 287, 287 (ill.), 346
Diesel, Rudolph 2: 368
Difference engine 2: 183, 184 (ill.)
Digestion, study of 1: 130
Diphtheria 1: 135, 136, 139, 150
Direct current (DC) 2: 325, 326
Dirichlet, Peter 2: 167
Dirigibles 2: 369
Disease, causes 1: 39
Disease, identification of 1: 124–28
Disquistiones Arithmeticae 2: 191
Dix, Dorothea 1: 151
Döbereiner, Johann 2: 229
Dohrn, Anton 1: 68
Donkin, Bryan 2: 309
Dorn, Friedrich 2: 244
Dubois, Marie Eugène 1: 50, 68
Dubois-Reymond, Emil 1: 68
Duchenne, Guillaume 1: 151
Dummer, Ernst 2: 167
Durand, Peter 2: 342
Duryea, Charles 2: 307
Duryea, James F. 2: 307
Dyes 1: 135
Dynamite 2: 303, 345, 346

E

Earth
    age of 2: 215, 251, 255
    geologic dating 1: 67
    rotation 2: 288
    structure of 2: 292
Earth sciences 2: 213
Eastman, George 2: 322, 368
Economics, mathematical study of 2: 202, 207
Edison, Thomas 2: 224, 302, **359–61**, 360 (ill.), 371
    light bulbs, incandescent 2: 326
    telegraph 2: 316
The Ego and the Id 1: 138
Ehrenberg, Christian 1: 68
Ehrlich, Paul 1: 81, **135–37**, 136 (ill.)
Eiffel, Gustave 2: 368
Einstein, Albert 2: 172, 226
Elasticity, law of 2: 233
Electric arc 2: 286
Electric current 2: 221

Index

Electric motors 2: 225
Electric shock therapy 1: 112
Electrical fields 2: 166
Electricity 2: 213, **324–30**
    ampere 2: 285
    environmental impact 2: 329
    magnetic effects 2: 213
    mathematical theory of electricity 2: **168–72**
Electrochemical dualism, theory of 2: 285
Electrochemistry 2: **255–61**
Electrolysis 2: 256
Electromagnetic fields 2: 221, 273, 274, 279, 283
Electromagnetic induction 2: 280
Electromagnetic theory 2: **221–26**
Electron shells 2: 244
Electrons 2: 292
Electrophysiology 1: 68
Electroplating 2: 260
Elementar Mathematik von Hoheren Standpunkte aus 2: 195
Elements 2: 227
    atomic theory of elements 2: 214
    atomic weight 2: 228
    periodic law of elements 2: 215
    periodic table of elements 2: **226–32**
Elements of Agricultural Chemistry 1: 18
Elements of Pure Economics 2: 207
Elephantiasis 1: 117, 118, 154
Elevators 2: 370
Elliptical functions 2: 166
Elson, Mary Anna 1: 151
Embryology, experimental 1: 72
Empedocles 1: 29
Encke, Johann 2: 287
Energy 2: 218
    development of the concept of energy 2: **216–21**
    law of conservation of energy 2: 217
    principle of conservation of energy 2: 289
Engines, internal combustion. *See* Internal combustion engines
Engines, steam 2: 330
    publishing industry 2: 308
An Enquiry Concerning Political Justice 1: 26
Enteric Fever 1: 117
Entomology 1: 72
Entrepreneurship and industrialization 2: 301
Entropy 2: 287
Environment, electricity and 2: 329
Enzymes 1: 71
Eole 2: 366
Epidemics 1: 117, 122
Ericsson, John 2: 368
Erosion 2: 287
Escherich, Theodor 1: 152
Escherichia coli 1: 152
Esmarch bandage 1: 114
Esmarch, Johannes Freidrich August von 1: 113
An Essay on the Principle of Population 1: 25
Essays on the Inequality of the Human Races 1: 99
Ether 1: 87
Euclid's fifth postulate 2: 198
Eugenics 1: 69
Euler, Leonhard 2: 167
Evolution 1: 29–36
    human evolution 1: **48–53**
    Darwin, Charles 1: 6
    Haeckel, Ernst 1: 58

Index

Huxley, Thomas 1: 59
Lamarck, Jean-Baptiste de 1: 70
Oken, Lorenz 1: 64
theory of evolution 1: 29–36, 66, 71, 2: 254
Wallace, Alfred 1: 65
Experimental psychology 1: 82
Experiments with Plant Hybrids 1: 22
Explosives 2: 344–48

F

Faraday, Michael 1: 46 (ill.), 47, 2: 169, 221, 259, 274 (ill.), 280, 314, **324–30**
    electric fields 2: 247, 273–75
Faraday's Law 2: 280
Father of
    antiseptic surgery 1: 93
    experimental pharmacology 1: 154
    geology 2: 282
    histology 1: 13
    mathematical logic 2: 185
    modern cryogenics 2: 274
    modern dentistry 1: 107
    modern geology 2: 254
    modern road building 2: 369
    symbolic logic 2: 185
    taxonomy 1: 8
    theory of evolution 1: 66
    tropical medicine 1: 118
    See also Founder of
Fauchard, Pierre 1: 107
Fermat, Pierre de 2: 179, 181 (ill.)
Fermat's last theorem
    Germain, Sophie 2: 194
Fermentation 2: 340, 341
Fertilizer 1: 17
Finite group theory 2: 206
Finsen, Niels Ryberg 1: 152
First 2: 174
    American research mathematician 2: 205
    automobile of modern design 2: 305
    battery 2: 255, 313, 324
    calculator, modern 2: 367
    commercial railroad system 2: 300
    computer, mechanical 2: 182
    computer program 2: 358
    director of the Geological Survey of Canada 2: 289
    electric battery 2: 169
    electric power station 2: 327
    facsimile 2: 366
    incandescent electric light system 2: 361
    internal combustion engine, four-stroke 2: 370
    internal combustion engine, usable 2: 369
    journal devoted entirely to mathematics 2: 202
    journal devoted to mathematics teaching 2: 174
    lighting system, coal gas 2: 370
    locomotive, steam 2: 371
    loom for weaving, automatic 2: 336
    machine gun, fully automatic 2: 369
    modern gasoline powered automobile 2: 354
    motorcycle 2: 306
    movie camera 2: 361
    non-German made an honorary professor at the University of Berlin 1: 140
    oceanographer 1: 62

Index

Ph.D. by University of Heidelberg to a woman *1:* 62
plastic *2:* 370
practical locomotive *2:* 330
president of the New York Academy of Medicine *1:* 156
privately produced mathematical journal *2:* 167
privately published journal of mathematics *2:* 174
professional mathematical society in the United States *2:* 167, 174
professional medical society in the United States *1:* 85
professional woman astronomer in the United States *2:* 290
professor of botany at Harvard University *1:* 69
public institution for the mentally retarded *1:* 105
railroad sleeping car *2:* 370
refrigerator, commercially successful *2:* 343
refrigerator, mechanical *2:* 367
revolver *2:* 367
school of osteopathy *1:* 87
secretary of the Smithsonian Institution *2:* 279
sewing machine *2:* 365, 369
steam powered locomotive *2:* 330
successful tracheotomy *1:* 150
table of atomic weights *2:* 272
telegraph *2:* 313
telegraph message *2:* 315
telephone *2:* 316
telephone directory *2:* 319
telephone switchboard system *2:* 319
telephones for commercial use *2:* 319
text on probability in English *2:* 180
textbook on internal medicine *1:* 149
textbook on psychology in English *1:* 149
transatlantic telephone cable *1:* 63
woman dentist in the United States *1:* 153
woman elected to the American Academy of Arts and Sciences *2:* 290
woman elected to the Institution of Electrical Engineers *2:* 286
woman elected to the Royal Society *2:* 358
woman in American history to earn a medical degree from a recognized medical college *1:* 132
woman invited to sessions at the Institute of France *2:* 194
woman listed in the Medical Register of the United Kingdom *1:* 135
woman permitted to read her paper before the Royal Society of London *2:* 286
woman to attend the meetings of the French Academy of Science *2:* 194
woman to be given an honorary doctorate *2:* 194
woman to earn a Ph.D. in Europe *2:* 269

Index

woman to hold a professorship at the Sorbonne *2:* 270
woman to hold and independent government clerkship, U.S. *1:* 128
woman to receive the Order of Merit *1:* 142
woman to win a Nobel Prize *2:* 269
women graduates of the Women's College of Philadelphia *1:* 151
First aid kit *1:* 114
Fish Commission, U.S. *1:* 58
Fizeau, Armand *2:* 287
Flemming, Walther *1:* 14, 69
Flora Americae Septentrionalis *1:* 72
Flourens, Marie-Jean *1:* 69
Food preservation *2:* 292, 301, **340–44**
Forbes, Edward *1:* 69
Ford, Henry *2:* 306 (ill.), 307
Formulary of Mathematics *2:* 177
Fossil fuel *2:* 325
Fossils *1:* 7, 70, *2:* 291
    dinosaurs *1:* 70
    human *1:* 50
Foucault, Jean *2:* 288
Foucault pendelum *2:* 288
Founder of
    experimental embryology *1:* 72
    field of abdominal surgery *1:* 150
    genetics *1:* 5
    hematology *1:* 135
    mathematical physics *2:* 193
    microbiology *1:* 42
    modern entomology *1:* 70
    modern mathematical logic *2:* 177
    non-Euclidean geometry *2:* 197
    nutrition science *1:* 20
    See also Father of
Fourdrinier, Henry *2:* 309
Fourdrinier, Sealy *2:* 309
Fourier, Jean-Baptiste *2:* **187–89**, 188 (ill.), 224
Fourier series *2:* 166, 187
Franklin, Benjamin *2:* 324
Free-cell formation, theory of *1:* 16
Frege, Gottlob *2:* 177
Frequency *2:* 221
Fresnel, Augustin *2:* 288
Freud, Anna *1:* 137
Freud, Sigmund *1:* 82, **137–39**, 138 (ill.)
Friedlander, Julius R. *1:* 105
Fulton, Robert *2:* 300
Fundamental theorem of algebra *2:* 191

G

Gale, Leonard *2:* 314
Galen *1:* 81
Galilei, Galileo *2:* 213, 233
Gall, Franz Joseph *1:* 152
Gallaudet, Thomas *1:* 103
Galois, Evariste *2:* 187, **189–90**, 189 (ill.)
Galton, Francis *1:* 69, 99, *2:* 182
Galvanocautery *1:* 155
Gamma globulins *1:* 151
Gas lantern *2:* 285
Gases, inert *2:* 243
Gases, kinetic theory of *2:* 292
Gaspart, Jean Marc *1:* 105
Gauss, Carl *2:* 165, **190–92**, 191 (ill.), 198
    Germain, Sophie *2:* 193

Gay-Lussac, Joseph 2: 228, 229, 275–77, 276 (ill.)
Gay-Lussac's Law 2: 276
Geissler, Henrich 2: 368
Geissler tube 2: 368
Gender theories 1: 100
Gene therapy 1: 22, 24
Generator, electrical 2: 221
Generators 2: 325
Genes 1: 22
   See also Genetics, Heredity, Mendel, Gregor
Genetics 1: 22
   Mendel, Gregor 1: 21–25
   Stevens, Nettie Maria 1: 73
   Weisman, August 1: 74
Geneva Medical College 1: 132
Geologic dating 1: 67
Geometry 2: 173
   algebraic geometry 2: 207
   descriptive geometry 2: 204
   Hilbert, David 2: 203
   non-Euclidean geometry 2: 165, 197
   projective geometry 2: 166, 194
Gergonne, Joseph Diaz 2: 167
Germ theory 1: 39–45, 81, 91, 2: 341
Germain, Sophie 2: 193–94, 193 (ill.)
Gibbs, Josiah 2: 277–79, 278 (ill.)
Giffard, Henri 2: 369
Gilbert, William 2: 221
Glacial drift 2: 262
Glaciers 2: 262, 265
Gobineau, Joseph Arthur de 1: 99
Godwin, Hannibal 2: 322
Godwin, William 1: 26
Goette, Alexander 1: 61
Gondwana 2: 292

Goodyear, Charles 2: 301, **348–52**, 350 (ill.)
Gorgas, William Crawford 1: 152
Gorrie, John 2: 343
Gram, Hans 1: 85
Gramme, Zenobe 2: 325
Grassman, Hermann 2: 203
Gray, Asa 1: 47, 69
Gray, Elisha 2: 317
Gray, Henry 1: 152
Gray's Anatomy 1: 152
Great Chain of Being 1: 59
Great Western Railway 2: 367
Green, George 2: 166, 169, 170
Greenhouse effect 2: 262, 264
Group sessions 1: 110
Group theory 2: 189
   finite group theory 2: 206
Guerin, Camille 1: 97
Gulf Stream 1: 63
Gum elastic 2: 349
Guncotton 2: 344, 345
Gunpowder 2: 344, 345
Gutenberg, Johannes 2: 310
Gutta percha 1: 108

H

Haeckel, Ernst 1: 6, **58–59**, 58 (ill.)
Hahnemann, Samuel 1: 84, 86
Hall, Charles M. 2: 260
Hamilton, Alice 1: 133, 133 (ill.)
Hancock, Thomas 2: 349, 351
Handicapped, schools for 1: **103–6**
Hansen, Armauer 1: 118
Hatch Agricultural Experiment Station Act of 1887 1: 19
Health, effects of colonization on 1: 117
Health, public. *See* Public health system

Index

Heat, mechanical and dynamic theory of 2: 288
Heine, Jakob von 1: 153
Heliotrope 2: 191
Helmholtz, Hermann 1: 6, 2: 223, 278, 288
   law of conservation of energy 2: 219
Henry, Joseph 2: 279–80, 279 (ill.), 314, 325
Henry, William 2: 288
Henry's Law 2: 288
Herbal medicines 1: 85
Heredity 1: 5, **21–25**
   theory of natural selection 1: 34
Herschel, Caroline 2: **280–82**, 281 (ill.)
Hertz, Heinrich 2: 171, 214, **247–51**
   radio waves 2: 224
Hickman, Henry Hill 1: 90
Hilbert, David 2: 203
   Pasch, Moritz 2: 205
Hill, George 2: 203
Hippocrates
   disease, causes of 1: 39
   human race 1: 8
Histology 1: 11, 13
H.M.S. *Beagle* 1: 32
Hodgkin, Thomas 1: 153
Hoe, Richard 2: 311
Hollerith, Herman 2: 338
Homeopathy 1: 84, 85
Homo erectus 1: 50
Homunculus 1: 21
Hooke, Robert 1: 10
Howe, Elias 2: 365, 369
Howe, Samuel 1: 105
Hubbard, Dardiner 2: 319
Huggins, Margaret 2: 288
Huggins, William 2: 288

Human race, biological theories of 1: **98–100**
Human race, classification of 1: **7–10**
Humors (and health) 1: 40
Hutton, James 2: 251, 283
Huxley, Thomas 1: **59–61**, 60 (ill.)
   Darwin, Charles 1: 60
Huygens, Christiaan 2: 179, 218
Hyde, Ida 1: **61–62**
Hydroelectric plants 2: 325
Hygrometer 2: 368
Hypercomplex numbers 2: 166
Hypodermic needle 1: 155
Hysteria 1: 137

I

IBM 2: 339
Ice ages 2: **261–64**, 265, 285
Idealism 1: 5
Iguanodon 1: 70
Immunization 1: 42
Index to Falmsteed's Observations of the Fixed Stars 2: 281
India rubber 2: 349
Inductive reasoning 2: 176, 177
Industrial Revolution 1: 95, 2: 332
   corporations 2: 334
   effect on public health 1: 120
   energy and 2: 220
   entrepreneurship, rise of 2: 301
   impact of mathematics on 2: 171, 173
   impact on literacy 2: 311
   impact on science 1: 47
   locomotives, steam powered 2: 330
   scientific management 2: 371
   stock market 2: 301

Index

women in the work place 2: 318
Inert gases 2: 243–47
Infinity 2: 166, 201
Infusoria 1: 64, 65
Inheritance of acquired characteristics 1: 30
Integral equations 2: 207
Internal combustion engines 2: 300, 303, 305, 369
    diesel 2: 368
    four-stroke engines 2: 370
International Business Machines (IBM) 2: 339
The Interpretation of Dreams 1: 138
An Investigation into the Laws of Thought 2: 185
Iron 2: 357

J

Jacobi, Mary Putnam 1: 102
Jacquard, Joseph-Marie 2: 336
Jacquard loom, reaction to 2: 339
Java Man 1: 50, 68
Jenner, Edward 1: 42
Jex-Blake, Sophia 1: 135, 153
Jones, Emeline 1: 153
Joule, James 2: 214, 218, 278, 289 (ill.)
Joule's Law 2: 289
Jung, Carl Gustav 1: 138

K

Kellner, Oskar 1: 20
Kelvin scale 2: 292
Kidney function 1: 150
Kinetic energy 2: 216
Kinetic theory of gases 2: 292
Kinetograph 2: 322, 361
Kinetoscope 2: 320, 322
Kirchoff, Gustav 2: 289
Kitasato, Shibasaburo 1: 81, 136, **139–40**

Klein, Felix 2: **194–95**
Knight, Charles 1: 52
Knight, Margaret 2: **361–62**
Koch, Robert 1: 81, 94, 122, 124–28, 128 (ill.), 139
Koch's Postulates 1: **124–28**
Kolliker, Rudolf 1: 146
König, Friedrich 2: 311
Kossel, Karl 1: 70
Kovalesvskaya, Sofia 2: 203
Kurdach, Karl 1: 3
Kussmaul, Adolf 1: 153

L

La Peyrere, Isaac de 1: 49
Lady Edison 2: 361
LaGrange, Joseph-Louis 2: 193, 197
Lagrange, Pierre 2: 186
Lamarck, Jean-Baptiste de 1: 3, 29, 70, 98
Landsteiner, Karl 1: 100
Langen, Eugen 2: 356
Langley, Samuel 2: 369
Lantern, gas 2: 285
Laplace, Pierre-Simon 2: 167, 174, 181, 186, **195–97**, 196 (ill.), 275
Large numbers, law of 2: 182
Larrey, Dominique-Jean 1: 113
Lartet, Édouard-Armand 1: 70
Latex 2: 349
Latreille, Pierre-André 1: 70
Laughing gas. *See* Nitrous oxide
Lavater, Johann 1: 112
Laveran, Charles 1: 81, **140–41**, 141 (ill.)
Lavoisier, Antoine 1: 17, 18 (ill.)
Law of conservation of energy 2: 214, 217, 219
Law of large numbers 2: 182
Law of triads 2: 227, 229

Index

Leeuwenhoek, Anton van *1:* 40
Legendre, Adrien-Marie *2:* 193
Leibniz, Gottfried *2:* 165, 176
Leishman, William Boog *1:* 118
Leishmaniasis *1:* 117, 118
Lemoine, Emile *2:* 204
Lenoir, Étienne *2:* 300, 369
   internal combustion engine *2:* 305
Leprosy *1:* 118
Letterman, Jonathan *1:* 114
Leverrier, Urbain *2:* 267, 285
Lewis, John *1:* 108
Lie, Sophus *2:* 206
Liebig, Justus von *1:* 18, 70, 74
Light bulb *2:* 326, 330 (ill.), 359, 371
Light waves *2:* 171, 293
Light
   diffraction *2:* 288
   scattering *2:* 292
   speed of *2:* 284, 287, 289
   wave theory of light *2:* 224
Linde, Karl von *2:* 343
Linnaeus, Carolus *1:* 8, 46
Linotype *2:* 309, 310
Liouville, Joseph *2:* 190
Lip reading *1:* 104
Lister, Joseph *1:* 82, 91
Listing, Johann *2:* 204
Literacy *2:* 311
Lobachevsky, Nickolai *2:* 165, **197–99**, 198 (ill.)
Locomotives, electric *2:* 371
Locomotives, steam powered *2:* 300, **331–35**, 371
   Pullman cars *2:* 335
Logan, William *2:* 289
Logic *2:* 202
   mathematics of logic *2:* **175–79**
   symbolic logic *2:* 176

Logopedics *1:* 154
London Medical College for Women *1:* 135
Long, Crawford *1:* 91, 154
Lumiére, Auguste *2:* 302, 322
Lumiére, Louis *2:* 302, 322
Lyell, Charles *2:* **251–55**, 282–83, 282 (ill.)

M

Mach, Ernst *2:* 289
Macintosh, Charles *2:* 349
Magendie, François *1:* 154
Maggot therapy *1:* 113, 114
Magnesium *2:* 258
Magnetic fields *2:* 166
   See also Electromagnetic fields
Magnetic resonance imaging (MRI) *2:* 248, 250
Magnetism *2:* 192, 221
   See also Electromagnetic fields
Malaria *1:* 118, 119
   transmission of by protozoa *1:* 140
Malta fever *1:* 150
Malthus, Thomas *1:* **25–29**, 25 (ill.)
Mammals of North American *1:* 58
Manson, Patrick *1:* 118, 154
Mantell, Gideon *1:* 7, 70
Marconi, Guglielmo *2:* 249
Massachusetts School for Idiotic and Feeble Minded Youth *1:* 105
Mast cells *1:* 135
Mathematical theory of electricity *2:* **168–72**
Mathematics *2:* 202
   astronomy *2:* 195
   automorphic functions *2:* 200
   education in mathematics *2:* **172–75**

Index

Fourier series 2: 187
golden age of mathematics 2: 165
group theory 2: 189
history 2: 201
industrial revolution 2: 171
integral equations 2: 207
logic 2: **175–79**, 201
n-body problem 2: 200
partial differential equations 2: 203
physics 2: 193
projective geometry 2: 194
string geometry 2: 201
symbolic logic 2: 185
theory of electricity 2: 167
theory of probability 2: 205
Matter, particulate theory of 2: 272
Maury, Matthew 1: **62–64**, 63 (ill.)
Maxim, Hiram 2: 369
Maxwell, James Clerk 2: 170, 206, 213, 247, 279, **283–84**, 284 (ill.)
electromagnetic theory 2: 221
light 2: 214
Mayer, Julius 2: 218
McAdam, John 2: 369
McCormick, Cyrus 2: **362–63**, 363 (ill.)
Mécanique Céleste 2: 196
Mechanical and dynamic theory of heat 2: 288
Mechanical reaper 2: 362
Mechanistic philosophy 1: 3
Medicine
aseptic techniques 1: 147
education in medicine 1: 143, 153, 155
military medicine 1: 113–16, 142
nurses and nursing 1: 115, 128, 141–42
tropical medicine 1: 116–20
See also Alternative Medicine; Surgery
Meiosis 1: 11, 15
Méliès, George 2: 324
Men of the Old Stone Age 1: 52
Mendel, Gregor 1: 5, **21–25**, 22 (ill.), 34
Mendeleyev, Dmitri 2: 229, 230–31
periodic table 2: **226–32**
Mental illness
hospitals for 1: 151
patients and their treatment 1: **109–13**, 144
Westphal, Carl Friedrich 1: 156
Mental retardation 1: 104
Mentally retarded, education of 1: **105–6**
Mercury pump 2: 367
Mergenthaler, Ottmar 2: 310
Metaphase 1: 15
Mette, Albert 1: 97
Meyer, Lothar 2: 230
Michelson, Albert 2: 216, 289
Microbes 1: 40, 125
See also Germs, Germ theory
Microbiology 1: 42
Micrographia 1: 11
Microorganisms 1: 40
Micropaleontology 1: 68
Microscopes and advances in cell theory 1: 10
Microscopical Researches into the Accordance in the Structure and Growth of Animals and Plants 1: 16
Microtome 1: 72
Microwave ovens 2: 250

Index

Mid-Atlantic Ridge *1:* 63
Miescher, Johann *1:* 71
Military medicine *1:* **113–16**
   Nightingale, Florence *1:* 141–42
Miner's safety lamp *2:* 258
Mitchell, Maria *2:* 290, 291 (ill.)
Mitosis *1:* 11, 15, 69, 73
Mobile army hospital *1:* 114
Möbius, August *2:* 202
Möbius strip *2:* 204
Mohl, Hugo von *1:* 14, 71
Moivre, Abraham de *2:* 180
Molecular weight *2:* 229
Molecules *2:* 277
Monge, Gaspard *2:* 204
Monitor *2:* 368
Montessori, Maria *1:* 105
Montessori schools *1:* 106
Moore, Eliakim *2:* 204
Morse, Samuel *2:* 280, 300, 314 (ill.)
   telegraph *2:* 314
Morton, William *1:* 92, 154
Mosquitoes, disease transmission and *1:* 118, 152, 155
Motion pictures *2:* **320–24**, 359
Motorcycle *2:* 306
Movies. *See* Motion pictures
MRI *2:* 248
Murdock, William *2:* 370
Muscular dystrophy *1:* 151
Museums, growth of *1:* 47

N

N-body problem *2:* 200
Nägeli, Karl *1:* 71
Naples Zoological Station *1:* 68
National Audubon Society *1:* 56
National Museum of Natural History *1:* 56
Natural history *1:* 46
Natural philosophy *1:* 5
Natural selection, theory of. *See* Theory of natural selection
Neanderthal man *1:* **48–53**
   misconceptions of *1:* 52
Nebulae *2:* 288
Nebulae, catalogs of *2:* 282
Nebular hypothesis *2:* 196
Neon *2:* 246
Neon light *2:* 243, 246
Neptune *2:* 267, 285
New England Asylum for the Blind *1:* 104
New System of Chemical Philosophy *2:* 273
New York Infirmary for Women and Children *1:* 134
New York Institute for the Deaf *1:* 104
New York Mathematical Society *2:* 167
New York point system *1:* 105
Newton, Isaac *1:* 4, *2:* 165, 180, 213, 218
Nicholson, William *2:* 256
Nickelodeons *2:* 323
Niemann, Albert *1:* 154
Niepce, Joseph *2:* 322
Nightingale, Florence *1:* 83, 134, **141–42**, 142 (ill.)
Nitroglycerine *2:* 345
Nitrous oxide *1:* 87, 90
Nobel, Alfred *2:* 303, 346 (ill.)
   establishment of Nobel Prize *2:* 346, 348
Nobel Prize in chemistry *2:* 271, 285
Nobel Prize in physics *2:* 240, 269, 270, 290
Nobel Prize in physiology or medicine *1:* 70, 140, 155
Noble gases *2:* 243

Index

Non-Euclidean geometry 2: 165
Normography 2: 203
Nuclei 1: 67
Nucleic acids 1: 71
Nucleoplasm 1: 73
Nucleus 1: 11, 14
Number theory 2: 167, 173, 191, 194
Nurses and nursing 1: 115
    Nightingale, Florence **1: 141–42**
    Barton, Clara **1: 128–29**
Nutrition science 1: 20
Nutt, Emma M. 2: 318

O

Ocean currents 1: 63
Oceanography 1: 62, 69
Oersted, Hans Christian 2: 169, 213, 218, 290, 314
    magnetic fields 2: 222, 247, 274
Office of Alternative Medicine, NIH 1: 89
Ohm, Georg Simon 2: 290
Ohm's Law 2: 290
Oken, Lorenz **1: 64–65**, 64 (ill.)
Olbers, Heinrich 2: 267, 291
Olbers's paradox 2: 291
Olds, Ransom 2: 307
On the Origin of Species 1: 29
On the Unity of Mankind 1: 7, 9
Optical telescope 2: 248
Organic Chemistry and Applications to Agriculture and Physiology 1: 18
Organon 2: 176
Orton, William 2: 319
Osborn, Henry 1: 52
Oscillating circuits 2: 248
Oscillation 2: 248

Osler, William **1: 143–44**, 143 (ill.)
Osteopathy 1: 84–86, 156
Otis, Elisha Graves 2: 370
Otto, Nicolaus 2: 300, 356, 370
Ozone 2: 291

P

Palladium 2: 292
Pallas 2: 291
Palmer, Daniel 1: 89
Paper 2: 308
Paper bags 2: 362
Parallax 2: 267
Paranoia 1: 156
Parkes, Alexander 2: 370
Parkesine 2: 370
Partial differential equations 2: 203
Particulate theory of matter 2: 272
Pascal, Blaise 2: 179
Pasch, Moritz 2: 205
Pasteur, Louis 1: 6, **39–45**, 39 (ill.), 81, 91, 122
Pasteurization 1: 41
Pavement 2: 369
Pavlov, Ivan Petrovich 1: 155
Payen, Anselme 1: 71
Peano, Giuseppe 2: 177
Peck, William 1: 72
Peirce, Benjamin 2: 205
Pelletier, Pierre-Joseph 1: 72
Pepper, William 1: 155
Pericardium 1: 114
Periodic law of elements 2: 215
Periodic motion 2: 187
Periodic table **2: 226–32**
Perraudin, Jean-Pierre 2: 262
Persistence of vision 2: 320
Pestalozzi, Johann 2: 175, 176 (ill.)
Pesticides 1: 17

Index

Pfeffer, Wilhelm *1:* 72
Phase rule *2:* 278
Phonograph *2:* 359, 361
Photography *2:* 293, 368, 371
Phrenology *1:* 110, 112, 152
Physiognomy *1:* 112
Pickling (foods) *2:* 340
Pinel, Philippe *1:* 82, **144–45**, 144 (ill.)
Pitchblend *2:* 270
Pithecanthropus *1:* 50, 68
Pixii, Hippolyte *2:* 325
Planck, Max *2:* 226
Plants, classification of *1:* 8
Plastic *2:* 370
Play therapy *1:* 110
Playfair, John *2:* 252
Pneumatic Medical Institute *1:* 90
Poincaré, Jules-Henri *2:* **199–200**, 199 (ill.)
Poisson, Siméon *2:* 205
Poliomyelitis *1:* 153
Polonium *2:* 237, 240, 269, 270
Population growth, theory of *1:* 28
Population studies *1:* **25–29**
Potential energy *2:* 169, 216, 217
Poultice *1:* 107
Pravax, Charles Gabriel *1:* 155
Preparing the Child for Science *2:* 201
Preservation of food. *See* Food preservation
Priestley, Joseph *1:* 90, *2:* 244, 349
Principle of conservation of energy *2:* 289
The Principles and Practice of Medicine *1:* 143
The Principles of Geology *2:* 251, 253
Printing press *2:* 311
Printing, carbon process of *2:* 371

Probability theory *2:* **179–82**, 205
Prodromus *1:* 68
Projective geometry *2:* 194
Prophase *1:* 15
Protoplasm *1:* 71
Psychiatry *1:* 110
Psychotherapy *1:* 145
    Freud, Sigmund *1:* **137–39**
Public Health Act of 1848 *1:* 122
Public health system *1:* **120–24**
    France *1:* 123, 124
    United States *1:* 124
Publishing industry *2:* **308–13**
Pullman cars *2:* 335, 370
Pullman, George *2:* 335, 370
Punched cards *2:* 299, 337
    computers and *2:* 338
Purkinje, Jan *1:* 72
Pursh, Frederick *1:* 72

Q

Quantum theory *2:* 225
The Question of Rest for Women During Menstruation *1:* 103
Quetelet, Adolphe *2:* 182, 182 (ill.)
Quintelet, Lambert *2:* 167, 181
Quintic equations *2:* 201

R

Race and racism *1:* 98, 99
Race, environmental theory of *1:* 98
Race, theories of *1:* **98–100**
Radio telescope *2:* 248, 250
Radio waves *2:* 171, 221, 224, **247–51**
Radioactivity *2:* **236–43**, 269
    medical applications *2:* 241
Radium *2:* 237, 240, 269, 270
Radium Institute *2:* 271
Raff, Norman *2:* 322

Index

Railroad brake 2: 277
Railroads
    Great Western Railway 2: 367
    locomotives, electric 2: 371
    locomotives, steam powered 2: 241, 300, 331–35, 371
    Pullman cars 2: 335
Raincoats 2: 349
Reaper, mechanical 2: 362
Recapitulation 1: 59
Red Cross
    American Red Cross 1: 128
    International Red Cross 1: 124
Reed, Walter 1: 119, 155
Refrigeration (foods) 2: 343
Refrigerators 2: 367
Reimann, Georg 2: 167
Reis, Johann 2: 316
Renault, Louis 2: 306
Retizius, Anders 1: 98
Reynaud, Emile 2: 320
Rhodium 2: 293
Riemann, Bernhard 2: 191
Rillieux, Norbert 2: 363–65
Robert, Nicholas 2: 309
Rock strata 2: 291
Roentgen, Wilhelm 1: 145–47, 146 (ill.)
    radio waves 2: 224
    X rays 2: 237
Roget, Peter 2: 320
Romanticism 1: 5
Ross, Ronald 1: 118, 120, 141
Rotary press 2: 311
Roux, Wilhelm 1: 72
Royal Astronomical Society 2: 185
Royal Institution of Great Britain 1: 47
Rubber, vulcanization 2: 348–52
Russ, John D. 1: 105

S

Sabine, Wallace 2: 235
Sabotage 2: 339
Salvarsan 1: 136
Sanders, Thomas 2: 319
Sanitariums 1: 96
Sanitary Movement 1: 122
Sanitary techniques. *See* Aseptic techniques
Sanitoriums 1: 95
Saturn 2: 285
Sayre, Louis Albert 1: 155
Schleiden, Matthias 1: 5, 10, 13, 15
Schonbein, Christian 2: 291, 344
Schultze, Max 1: 73
Schwann, Theodor 1: 5, 10, 15
Science, popularization of 1: 45–48
Scientific anthropology 1: 9
Scientific botany 1: 13
Scientific management 2: 371
Scopes, John 1: 35
Scopes trial 1: 35
Screw propeller 2: 368
Second law of thermodynamics 2: 285
Séguin, Édouard 1: 105
Selden, George 2: 306
Selenium 2: 285
Semaphore 2: 313
Semmelweis, Ignaz 1: 82, 93, 147–49, 148 (ill.)
Set theory 2: 166, 173
    Venn diagram 2: 206
Seuss, Eduard 2: 292
Sewing machine 2: 365
Sex in Education: or, A Fair Chance for the Girls 1: 102
Sexias, David 1: 104
Siemens, Ernest von 2: 325, 370
Sign language 1: 103, 104

Index

Silliman, Benjamin 2: 291
Siméon, Denis 2: 205
Singer, Isaac 2: 301, **365–66**
Sith, William 2: 291
61 Cygni 2: 267, 268
Skin disorders 1: 152
Sleeping sickness 1: 126, 136
    transmission of 1: 151
Smith, Joel W. 1: 105
Smithsonian Institution 2: 280
    Baird, Spencer 1: 56
Snow, John 1: 122
Sobero, Ascanio 2: 345
Social Darwinism 1: **36–39**
Social sciences, statistics and 2: 167
Sodium 2: 258, 259
Sodium carbonate 2: 291
Soldiers, medical treatment 1: 113
Solvay, Ernest 2: 291
Somerville, Mary 2: 205, 358
Sonar 2: 233, 236
Sound, speed of 2: 289
Sound, theory of 2: **232–36**
Special education classes 1: 106
Special needs, establishment of schools for 1: **103–6**
Spectroscopy 2: 285
Spencer, Herbert 1: 36
Spurzheim, Johann 1: 112
St. Martin, Alexis 1: 130
Stable matter 2: 237
Statistical Society of London 2: 185
Statistics 2: 180
Statue of Liberty 2: 368
Staudt, Karl von 2: 206
Steam boiler 2: 304
Steamboats 2: 300
Steamships, screw propeller for 2: 368
Stearns, John 1: 156

Steel 2: 356
Stephenson, George 2: 330, 371
Stevens, Nettie Maria 1: 73
Still, Andrew 1: 86, 156
Stock market 2: 301
Strasburger, Eduard 1: 73
Streptomycin 1: 98
String geometry 2: 201
Strontium 2: 258
Strutt, John 2: **232–36**, 244
Sturm, Jacques 2: 206
Sugar refining 2: 363
Sumner, William 1: 37 (ill.), 38
Surgery
    advances in surgical techniques 1: **90–94**
    analgesia and anesthesia 1: 82, 90, 92
    arterial surgery 1: 154
    aseptic techniques 1: 82
    orthopedic 1: 155
    See also Military medicine
Survival of the fittest 1: 30
    Spencer, Herbert 1: 37
Swan, Joseph 2: 326, 371
Switchboard operators 2: 318
Syllogism 2: 176
Sylow, Peter 2: 206
Symbolic logic 2: 176, 177, 185
Syphilis, treatment of 1: 137

T

Table of atomic weights 2: 272
Tabulating Machine Company 2: 339
Tait, Peter 2: 206
Talbot, Henry 2: 371
Taveau, Auguste 1: 108
Taylor, Frederick 2: 371
TB. *See* Tuberculosis
Teeth, treatment of 1: 107

Index

Telegraph *2:* 249, 280, 300, 313–16
Telephone *2:* 301, **316–20**
    Bell, Alexander Graham *2:* 353
    Gray, Elisha *2:* 316
    switchboard operators *2:* 318
Telophase *1:* 15
Temperature *2:* 292
Tesla, Nikola *2:* 324, 326, 371
Tetansu *1:* 139
Textile weaving, mechanization of *2:* **336–40**
Thaumatrope *2:* 320
Thénard, Louis *2:* 276
Theory of electrochemical dualism *2:* 285
Theory of evolution *1:* 71, *2:* 254
    Darwin, Charles *1:* 6
    Haeckel, Ernst *1:* 58
    Huxley, Thomas *1:* 59
    Lamarck, Jean-Baptiste de *1:* 70
    Oken, Lorenz *1:* 64
    Wallace, Alfred *1:* 65
Theory of free-cell formation *1:* 16
Theory of natural selection *1:* 30, 32
    Malthus, Thomas *1:* 28
    Wallace, Alfred *1:* 33
Theory of population growth *1:* 28
Theory of probability *2:* 205
Theory of recapitulation *1:* 59
Theory of relativity *2:* 172, 191
    Maxwell, James Clerk *2:* 226
Theory of sets *2:* 166
Thermodynamics, chemical *2:* 277
Thermodynamics, second law of *2:* 285
Thermos bottle *2:* 287
Thimonnier, Barthélemy *2:* 365
Thomas, Charles *1:* 73

Thomas, William *2:* 369
Thompson, Benjamin *2:* 218
Thompson, Silvanius *2:* 240
Thomson, Joseph *2:* 292
Thomson, William *2:* 169, 171, 206, 292
Thorium *2:* 285
Till *2:* 262
Toplis, John *2:* 170
Topology *2:* 204
Tracers *2:* 237, 241
Transformers *2:* 372
Transportation, impact of railroads on *2:* 330
Trephination *1:* 87, 90
Trevithick, Richard *2:* 300, 330
Triage *1:* 114
    See also Military medicine
Tropical medicine *1:* **116–20**, 154
Trudeau, Edward *1:* 96
Trypanosomiasis *1:* 136
Tschermak, Erich *1:* 22
Tsetse fly *1:* 150
Tuberculosis *1:* **94–98**, 125, 135
    culture of *1:* 126
    transmission of *1:* 156
Tuke, William *1:* 112
Tylor, Edward *1:* 73
Tyndall, John *2:* 292
Type (printing) *2:* 308, 310

U

Ultrasound *2:* 233, 236
Uniformitarianism, principle of *2:* **251–55**, 283
United States Fish Commission *1:* 58
Uranium *2:* 238
Uranus *2:* 267, 268, 281
Urbanization, public health and *1:* 120
Urea *1:* 74

Index

V

Vaccines and vaccination *1:* 42, 95, 139
Vail, Alfred *2:* 314
Van Leeuwenhoek, Anton *1:* 21
Vanderbilt, Cornelius *2:* 334
Vaucanson, Jacques de *2:* 336
Vectors, calculus of *2:* 203
Venetz, Ignatz *2:* 262, 285
Venn, John *2:* 206
Villemin, Jean Antoine *1:* 156
Villermé, Louis-René *1:* 123
Viruses *1:* 44
"Vis viva" *2:* 217, 218
The Viviparous Quadrupeds of North America *1:* 56
Volta, Alessandro *2:* 167, 213, 218, 255, 313, 324
Voltaic pile *2:* 169, 256, 324
Volterra, Vito *2:* 207
Vorlesungen Über Geschichte der Mathematik *2:* 201
Vulcanization *2:* **348–52**

W

Wait, William B. *1:* 105
Waksman, Selman *1:* 98
Wallace, Alfred *1:* 10, 25, 65–66, 66 (ill.)
    uniformitarianism *2:* 254
Walras, Léon *2:* 207
War, medicine in. *See* Military medicine
Ward, Lester *1:* 38
Water Birds of North America *1:* 58
Waterproofing *2:* 349
Waterston, John *2:* 292
Watson, Thomas *2:* 318
Watt, James *2:* 330
Wave phenomena, analysis of *2:* 166
Wave theory *2:* 288
Weaving, mechanization of *2:* 336–40
Weierstrass, Karl *2:* 166
Weisman, August *1:* 74
Wells, Horace *1:* 91
Westinghouse, George *2:* 326
Westphal, Carl *1:* 156
What the Social Classes Owe to Each Other *1:* 38
Wilberforce, Samuel *1:* 60
Wilbur, Harvey B. *1:* 105
The Wild Boy of Aveyron *1:* 105
William, Crookes *1:* 146
Wöhler, Friedrich *1:* 74
Wollaston, William *2:* 292
Women inventors *2:* 361
Women, medical beliefs about *1:* **100–3**
Woods, Granville T. *2:* 372
World Health Organization *1:* 120
World War I, explosives and *2:* 347
Wrought iron *2:* 357
Wunderlich, Carl *1:* 156

X

X rays *1:* 145, *2:* 237

Y

Yellow fever *1:* 117, 118
Yersin, Alexander *1:* 140, 157
Young, Grace *2:* 207
Young, Thomas *2:* 218, 288, 293

Z

Zakrzewska, Maria *1:* 134, 157
Zeotrope *2:* 320
Zeppelin, Ferdinand von *2:* 369
Zeuthen, Hieronymous *2:* 207